GEOGRAPHY FOR LIFE ACTIVITIES
WITH ANSWER KEY
by Robin Elisabeth Datel

HOLT
PEOPLE, PLACES, AND CHANGE
An Introduction to World Studies

HOLT, RINEHART AND WINSTON

A Harcourt Education Company

Austin · Orlando · Chicago · New York · Toronto · London · San Diego

The Author

Robin Elisabeth Datel is an instructor in the Department of Geography at California State University, Sacramento. She received her B.A. from the University of California, Davis, and her M.A. and Ph.D. from the University of Minnesota, all in geography. She has written numerous articles on historic preservation and urban geography in the United States and Europe. She has also received grants from the American Association of University Women and the American Council of Learned Societies. Ms. Datel is a past president of the Association of Pacific Coast Geographers and has many years of experience teaching courses that introduce future K-12 teachers to geography.

Illustrations

All work, unless otherwise noted, contributed by Holt, Rinehart and Winston.

Chapter 1: Page 3, MapQuest.com, Inc.

Chapter 5: Page 15, MapQuest.com, Inc.

Chapter 6: Page 18, Leslie Kell.

Chapter 7: Page 19, MapQuest.com, Inc.

Chapter 10: Page 30, Ortelius Design.

Chapter 13: Page 39, Ortelius Design.

Chapter 14: Page 42, MapQuest.com, Inc.

Chapter 15: Page 45, Ortelius Design.

Chapter 16: Page 48, MapQuest.com, Inc.

Chapter 17: Page 51, MapQuest.com, Inc.

Chapter 18: Page 54, MapQuest.com, Inc.

Chapter 19: Page 57, MapQuest.com, Inc.

Chapter 23: Page 69, MapQuest.com, Inc.

Chapter 26: Page 77, MapQuest.com, Inc.

Chapter 27: Page 80, Leslie Kell.

Chapter 29: Page 87, Leslie Kell.

Chapter 30: Page 89, MapQuest.com, Inc.

Chapter 31: Page 93, Leslie Kell.

Chapter 32: Page 96, MapQuest.com, Inc.

Answer Keys: Page 107, MapQuest.com, Inc.; 109, MapQuest.com, Inc.; 111, MapQuest.com, Inc.; 112, MapQuest.com, Inc.; 116, MapQuest.com, Inc.

Cover: © Stone/Keren Su

Front and Back Cover Background, and Title Page: Artwork by Nio Graphics. Rendering based on photo by Stone/Cosmo Condina.

Printed in the United States of America

ISBN 0-03-066712-7

3 4 5 6 7 8 9 082 04 03

TABLE OF CONTENTS

Spatial Perspective

One of the most important ideas in geography is the idea of spatial perspective. A spatial perspective is a point of view that focuses on where things are located and why they are located there. This approach to studying the world is often used in geography. Geographers use a spatial perspective when they study Earth's landscapes, environments, and people.

In this exercise you will use a spatial perspective to analyze rates of deforestation in South America. Currently, about half of South America is covered with forests. Brazil has the largest area of forests. Brazil's Amazon Basin covers an area of approximately 1.5 million square miles (3.9 million square kilometers). However, since the 1980s rapid deforestation throughout South America has become a worldwide concern. Many people are worried that deforestation is damaging the environment, destroying entire species of plants and animals, and possibly even affecting Earth's climate. In the Andes, some slopes have been so severely deforested that it is hard to tell they were once covered with trees. Follow the steps below to apply a spatial perspective to deforestation in South America. Use the data from the table on page 2 and the map on page 3 to complete the activity.

YOU ARE THE GEOGRAPHER

1. Look at the table showing deforestation in South America between 1980 and 1990. What information does it contain? Which countries in South America had the highest rates of deforestation between 1980 and 1990? Which countries had the lowest rates of deforestation?

_____

_____

_____

2. Use the map of South America to apply a spatial perspective to the information in the table. At the top of the map, write the title *Deforestation in South America, 1980–1990.* Create a legend for your map. In the legend, draw five small boxes. Label these boxes as follows:

      0–5%

      5–10%

    10–15%

    15–20%

    20–25%

Now use a set of felt tip pens to color in the boxes. Leave the 0–5% box white, color the 5–10% box light gray, the 10–15% box red, the 15–20% box purple, and the 20–25% box black. Then use the information in the table and an atlas to color each South American country the appropriate color based on the amount of forest lost between 1980 and 1990.

3. How is your map different from the table? How does the map you created use a spatial perspective to show information? What patterns can you see on the map?

_____

_____

_____

4. The basic map that you created gives only a very general picture of deforestation in South America. It shows where deforestation occurred in the 1980s, but it does not explain why it occurred. Imagine you are a geographer getting ready to do a detailed study of deforestation in South America. What additional information would you need? Make a short list of some of this additional information. For example, you might want to see more detailed maps of deforestation within each country. You might also want to see maps showing population growth, logging, farming, and migration.

_____

_____

_____

_____

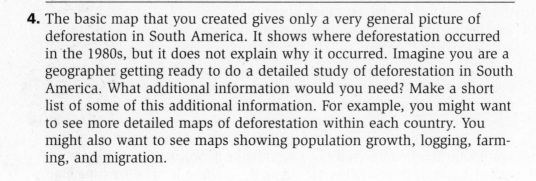

Deforestation in South America, 1980–1990

Country	Amount of Forest Lost
Argentina	5.9%
Bolivia	11.2%
Brazil	6.1%
Chile	7.9%
Colombia	6.4%
Ecuador	16.6%
Guyana	1.0%
Paraguay	23.8%
Peru	3.8%
Suriname	0.9%
Uruguay	1.5%
Venezuela	11.6%

ATLANTIC
OCEAN

PACIFIC
OCEAN

Hurricane Season

Hurricanes are large tropical storms that form over warm ocean waters. Hurricanes have wind speeds greater than 75 miles (121 km) per hour, cause heavy rains and thunderstorms, and can be hundreds of miles across. These large, powerful storms occur between the latitudes of approximately 5° and 30° in both the Northern and Southern Hemispheres. However, they are called hurricanes only in North America. In Asia and the western Pacific, they are called typhoons or cyclones. Scientists who study hurricanes try to predict when and where these dangerous storms will form so they can warn people who might be in the hurricane's path.

In this activity you will analyze the occurrence of some deadly Atlantic hurricanes. In North America, hurricanes form at a certain time of the year. You will identify this hurricane "season" by using the information in the table on page 5 to make a graph. Then you will analyze the relationship between the hurricane season in North America and the tilt of Earth on its axis and the effect of this tilt on temperatures. Follow the steps below to complete the exercise.

YOU ARE THE GEOGRAPHER

1. Look at the information in the table showing the names and approximate dates of some deadly Atlantic hurricanes. These hurricanes represent a sampling of some of the deadliest hurricanes that occurred in North America between 1950 and 2000.

2. Use the information in the table to make a graph. On page 6 the graph has been started for you. These two lines on page 6 will be the two sides of your graph. Just below the bottom line, label the 12 months of the year from left to right. You can use the abbreviations Jan., Feb., Mar., Apr., May, Jun., Jul., Aug., Sep., Oct., Nov., and Dec. Just to the left of the other line, write the numbers 1–10 from the bottom to the top. Now use the information in the table to count the total number of hurricanes that occurred in each month. Show this information on your graph by drawing a line above the months that had hurricanes up to the appropriate number. For example, three hurricanes occurred in October, so you would draw a line above the month of October until it is even with the number 3. When you are done, write the title *Some Deadly Atlantic Hurricanes, 1950–2000* at the top of the graph.

3. Analyze your graph. In which months did hurricanes occur? Of these, which month had the greatest number of hurricanes? In which months were there no hurricanes? Based on your graph, when do you think hurricane season in North America is?

_____

_____

4. Find a diagram in your textbook or in an atlas showing how Earth is tilted on its axis and revolves around the Sun, causing seasons. Compare the timing of the hurricane season in North America to the information in the diagram. What do you notice? During which seasons do hurricanes occur? Why do you think that is? What effect do you think Earth's tilt is having on ocean waters at this time?

_____

_____

_____

5. In the Southern Hemisphere, the seasons are reversed. For example, when it is summer in the Northern Hemisphere between June and September, it is winter in the Southern Hemisphere. Based on this fact and on the graph you created, when do you think hurricane season in the Southern Hemisphere is? Explain your answer.

_____

_____

Some Deadly Atlantic Hurricanes, 1950–2000

Hurricane Name	Approximate Dates
Agnes	June 19–22, 1972
Alberto	July 4–7, 1994
Allen	August 4–7, 1980
Alma	June 4–8, 1966
Andrew	August 23–27, 1992
Betsy	September 7–10, 1965
Bret	August 7–11, 1993
Cesar	July 25–28, 1996
Charlie	August 15–20, 1951
Cleo	August 22–26, 1964
Dora	September 12, 1956
Diane	August 16–19, 1955
Donna	September 4–5, 1960
Fifi	September 14–19, 1974
Gilbert	September 9–14, 1988
Gordon	November 8–21, 1994
Hazel	October 5–13, 1954
Hattie	October 26–31, 1961
Hilda	September 11–16, 1955
Hugo	September 17–22, 1989
Janet	September 22–28, 1955
Joan	October 14–22, 1988

Ocean Currents

Geographers try to explain and predict how Earth's physical processes shape patterns in the physical environment. The major ocean currents and wind belts that move across Earth are important in geography. Understanding these ocean currents and wind patterns is important to understanding weather and climate. In this activity you will map and predict how ocean currents and wind move ocean debris.

Major surface currents in the oceans move in large, slow circles called gyres. Gyres are produced by the wind around areas of high pressure. They occur in nearly the same position as the areas of high pressure over the ocean. Like the wind, these ocean currents flow clockwise in the Northern Hemisphere. What direction do you think they flow in the Southern Hemisphere?

The pattern of the world's major ocean currents helps explain the journey some Nike sneakers took through the Pacific Ocean. The container ship *Hansa Carrier* was in the northeastern Pacific Ocean on its way to the United States from Korea. On May 27, 1990, nearly 80,000 Nike sneakers went overboard. Six months to a year later, the shoes began to wash up on the shores of Washington, Oregon, and British Columbia. After washing the shoes and having the barnacles and oil removed from them, they were still wearable! Since the incident, scientists have constructed several computer models to understand the shoes' route. Now it is your turn. Use the data on page 9 and follow the steps below to complete the activity.

YOU ARE THE GEOGRAPHER

1. Find a map showing the world's ocean currents and winds. Describe what the map shows about the relationship between the world's ocean currents and winds.

_____

_____

_____

2. Look at the table on page 9. For how long did the shoes float in the ocean?

_____

Have all the shoes been recovered? If not, where do you think the others may be?

_____

_____

_____

3. Use a blank world map to trace the path of the shoes. Label each location with the numbers 1–11. Can you identify the nearest city, state, or country of these locations?

_____

_____

_____

_____

4. Which winds prevail in these areas?

_____

_____

After you have labeled the locations, follow the path of the shoes. Draw an arrow from the first date to the last. What shape is created?

_____

Did the shoes drift in a clockwise or counterclockwise direction?

_____

What term explains the pattern you drew?

_____

5. Where did the shoes land in 1996? Did the shoes complete the gyre's cycle?

_____

How long did it take for the shoes to come back?

_____

What additional factors might affect the route of the shoes?

_____

_____

_____

6. Although the Nike shoes did not seriously harm the environment, there are other types of ocean spills that do. Oil tankers sometimes spill their cargo. Imagine there was an oil spill in the same location where the _Hansa Carrier_ lost its cargo. Where would the oil go?

_____

_____

_____

What could happen to sea life?

_____

_____

_____

Think of the food chain around a coastline. How would the plants and animals be affected?

_____

_____

_____

Hansa Carrier Study

Location Number	Approximate Location	Number of Shoes Found	Date
1	161°W; 48°N	original shoe spill	May 27, 1990
2	135°W; 58°N	250	Mar. 26, 1991
3	130°W; 55°N	200	May 18, 1991
4	128°W; 52°N	100	Jan.–Feb. 1991
5	125°W; 49°N	200	Nov.–Dec. 1990
6	125°W; 48°N	200	Feb.–March 1991
7	124°W; 45°N	150	April 4, 1991
8	124°W; 42°N	200	May 9–10, 1991
9	156°W; 20°N	several	Jan.–Mar. 1993
10	120°E; 15°N	several	Jan.–July 1994
11	124°W; 45°N	several	Mar.–Apr. 1996

Desertification

GEOGRAPHY FOR LIFE ACTIVITIES

NATIONAL GEOGRAPHY STANDARD 14

Have you ever heard of desertification? Desertification is the spread of desert areas. It is caused by changing environmental conditions or unwise land use. Although the term was not used until almost 1950, desertification became well known in the 1930s when parts of the U.S. Great Plains turned into dusty, dry land because of drought and poor farming practices. You may have even heard the area referred to as the Dust Bowl. Many farmers lost everything during this time. Fortunately, the Great Plains recovered. The same cannot be said about some other places around the world.

WHAT CAUSES DESERTIFICATION?

Overfarming, overgrazing, overirrigation, and firewood gathering can all cause desertification. When desertification happens, plants disappear and the soil is eroded by wind and water. In some areas, people disrupt the local ecosystem and increase the rate of erosion. For example, desert nomads often herd animals. By pounding the soil with their hooves, animals like cattle compress the ground and cause erosion. In some areas, off-road vehicles cause erosion.

Many dry areas around the world suffer from desertification. These areas include North Africa, the "horn" of East Africa, regions around the Kalahari Desert, large parts of Southwest Asia, parts of Australia, eastern Brazil, southern Argentina, the central Andes, northern Mexico, and the southwestern United States. These dry lands are also home to about one sixth of the world's population. In some of these areas, annual population growth is 2.5–3.5 percent or more. Even in the most difficult drought year, the birthrate is often 10 times the death rate. This growing population means a higher demand for food and fuel, which in turn can lead to more desertification.

Worldwide about 2 billion acres of land have become desertified in the last 100 years. This is an area approximately the size of the United States. Each year about 40 million acres are added to this total.

YOU ARE THE GEOGRAPHER

1. The table on page 12 shows some cities that are subject to desertification. What is similar about these cities?

_____

_____

2. In your textbook or in an atlas, find the locations listed in the table. You may notice that most of these cities are not located in the desert. Instead, they are located on the edge of a desert and have a different type of vegetation. How is the regional environment classified in each of these cities?

_____

Explain how this affects the nearby desert.

Why are these cities are at risk of becoming desertified?

3. Because dry or semiarid lands are most subject to desertification and often border deserts, it sometimes seems like the desert "invades" the other land. However, this is not really true. It appears to happen this way because human-created deserts eventually look just like natural deserts. There is a gradual transition from a drier to a more humid environment, making it more difficult to define the desert border. These transition zones are vulnerable ecosystems. The process of desertification is not easy to map because it does not happen in a straight line. On a separate sheet of paper, draw a diagram showing the process of desertification.

4. Very often little or no data are available to show the rate of desertification. What approach would you use to study rates of desertification?

What evidence might you use?

How could you apply what you have learned to develop a plan to stop desertification in one of these regions?

Some Cities Subject to Desertification

City	Regional Population Density Inhabitants per sq. mile	Regional Environment	Vegetation	Climate
San Diego, California	10–25	Urban	Mediterranean	Semiarid
Hargeysa, Somalia	1–10	Grassland, grazing land	Savanna	Semiarid
Floriano, Brazil	1–10	Grassland, grazing land	Savanna	Semiarid
Monterrey, Mexico	10–25	Cropland, grazing land	Desert	Semiarid
Baghdad, Iraq	10–25	Cropland	Temperate grassland	Semiarid

What Region Do You Live In?

Geographers are interested in all kinds of regions. The regions they study are usually ones that they have defined and mapped after gathering data. However, some geographers are interested in another kind of region called the vernacular region. *Vernacular* means something that has been created by ordinary people. We could also call them popular regions, as popular means "of the people."

In 1980 geographer Wilbur Zelinsky created a map of the popular regions of North America (see page 15). This map shows in which region Americans and Canadians think their cities belong. Professor Zelinsky decided that the names ordinary people give their businesses or organizations could be used to determine the city's region. He looked at phone books for the 276 largest urban areas in the United States and Canada. He came up with a list of 73 terms that appeared frequently in the names of businesses and organizations (Table 1 on page 14 shows some of the terms). Some of these terms are regional and some are not. For each city, he counted the number of businesses that used each of the terms on his list in its name. The term used most frequently in business names determined the region to which that city was assigned. For example, there were more businesses in Oklahoma City, Oklahoma, and Lawton, Oklahoma, with the term *Southwest* in their names than any other, so those cities were assigned to the Southwest region. In Tulsa, OK, however, there were more businesses with the term *Midwest* in their names than any other, so Tulsa was considered part of the Midwest.

A couple of things about the map are worth pointing out. Two areas were impossible to assign to one region or the other, so the map shows them as belonging to both. One area is part of both the Northwest and the West and another is part of both the Southwest and the West. Also notice that there is a region on the map labeled as *No Regional Affiliation*. This region is made up of "metropolises in which no single regional term or cluster of terms occurs five or more times."

YOU ARE THE GEOGRAPHER

Now it is time to find out to which region your town belongs. Before you begin, decide which region *you* think it is in. Then obtain a copy of the local telephone book. Use the business pages. Count up how many businesses use each of the terms on Professor Zelinsky's list. Follow the same rules he did: do not count government agencies, do not count enterprises named for a person (thus, do not count a company named "Robert West Tires" under "West"), do not count businesses named for a local street (thus, do not count Western Day Care Center under "Western" if it is located on Western Avenue), and only count a business with many separate locations once (if there are five Southern Drive-Ins in your town, count that as one business).

1. In which region is your town?

_____

_____

Professor Zelinsky also looked at where the line between the perceived East and the perceived West lies in North America. You can do this easily by comparing the number of businesses in your locality using the term "West" (or a related term) to the number using the term "East" (or a related term).

2. Why do you think that the boundary between East and West is so far east?

_____

_____

You may want to do this exercise for several different places. You can look for telephone directories in your local library or on the Internet. Large metropolitan areas often have more than one phone book. Professor Zelinsky used only phone books for central cities, not suburbs.

Have a group discussion about this exercise. What does the regional label that applies to your city mean to you? What do you think of when someone uses a regional label? What kind of climate the region has? How the region looks? What the people there look like? What the people do for a living? How they speak? What they eat? What churches they attend? How well do you think popular (perceived) regions, in this case as defined by people picking names for their businesses and organizations, correspond with actual differences among regions?

Table 1   Terms Counted in the Study of Names of Metropolitan Businesses

Acadia	Metropolitan	Northwest(ern)
American	Mid-America(n)	Pacific
Apache	Mid-Atlantic, Middle Atlantic	Pilgrim
Atlantic	Midland	Pioneer
Canada, Canadian	Midway	Southeast(ern)
Colonial, Colony	Middle West, Midwest(ern)	Southern, Southland
Crown	National, Nationwide	Southwest(ern)
Dixie	New England	United States, U.S.
Eagle	Northeast(ern)	Victoria
Eastern	Northern	Viking
Federal	Northland	Western
Gulf	North Star	

Popular Regions of North America

Shopping Rules!

Urban geographers study cities. One topic they study is how land use is arranged in cities. Where are the manufacturing plants? Where do people live? How do shop owners decide where to put their stores? How does the transportation system affect where different types of activities are located?

This activity looks at one kind of land use, retailing, better known as shopping. Geographers might classify retail areas into three categories: centers, ribbons, and specialized areas. Ribbons are commercial districts strung out along a street or highway. Specialized areas are created when businesses selling the same product, like cars or antiques, are grouped together. Centers are also groups of retail businesses, but they contain a variety of stores. This activity is concerned with centers.

Shopping centers come in a variety of sizes. The largest store in a shopping center is called the anchor. The anchor in small shopping centers can be a grocery store, a drug/variety store, or a small department store. In bigger centers, full-size department stores are the anchors. Bigger centers are called either regional or superregional centers. Regional and superregional shopping centers are often referred to as "malls."

YOU ARE THE GEOGRAPHER

On a street map of your town or city, locate and mark the shopping centers. On your map, distinguish between (use different colors or symbols for) small and medium-sized centers on the one hand and regional and superregional centers on the other. If you live in a large city, you may be able to map the shopping centers in only part of the city, or only the regional and superregional centers throughout. Study the map.

1. How many small and medium-sized centers did you map? How many regional or superregional? About how far apart are the small and medium-sized centers?

_____

2. Are the regional and superregional centers in your city surrounded mostly by residential land uses (houses and apartments) or are they part of a larger cluster of commercial land uses (offices, other retailers, hotels, and so on)?

_____

_____

We can also study the internal geography of shopping centers. In a regional or superregional mall, the anchors are placed at opposite ends of the mall. Customers must then pass all the other shops to get from one anchor to the other. As for how to arrange the other shops, there are several plans. One plan groups together all the stores selling the same thing. Another plan spreads them out. Grouping the stores allows customers to comparison

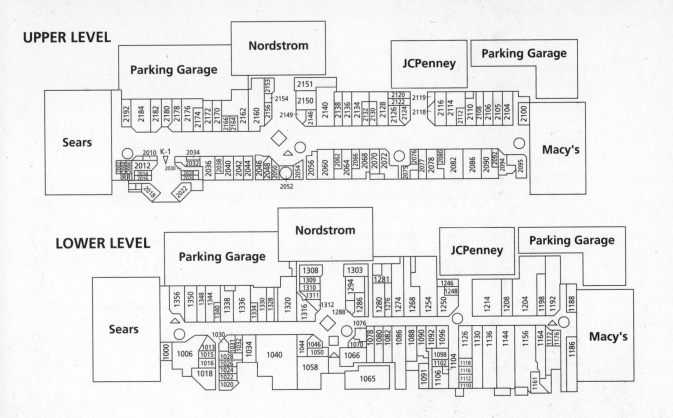

Type of Store	Color	Location
Apparel—Family and Children's	Red	1016, 1136, 1156, 1208, 1268, 1320, 2054, 2062, 2064, 2066, 2077, 2114
Apparel—Women's	Pink	1066, 1078, 1092, 1126, 1186, 1192, 1204, 1286, 1294, 1326, 1336, 2036, 2042, 2090, 2140, 2146, 2154, 2153, 2160, 2192
Apparel—Men's	Blue	1034, 1172, 2086, 2128, 2170, 2180
Beauty/Fitness and Health	Purple	1020, 1026, 1040, 1288, 1311, 1312, 2048, 2108, 2138
Electronics/Entertainment	Light green	1356, 2004, 2092, 2104, 2105, 2119, 2178, 2182
Home Furnishings and Accessories	Gray	1050, 1082, 1254, 2068, 2080, 2106, 2122, 2184
Jewelry	Orange	1013, 1032, 1044, 1076, 1096, 1276, 2076, 2126, 2149, 2156, 2164, 2166
Services	Leave white	1000, 1022, 1024, 1098, 1281, 1309, 1344, 2100, 2120
Shoes	Brown	1080, 1198, 1280, 1334, 1350, 2040, 2044, 2050, 2072, 2110, 2112, 2118, 2136, 2150, 2172
Specialty (includes books, cards, gifts, toys)	Black	1015, 1028, 1030, 1088, 1090, 1102, 1144, 1188, 1214, 1234, 1246, 1250, 1274, 1303, 1310, 1338, 1340, 2006, 2034, 2046, 2060, 2095, 2130, 2132, 2174, 2176
Sports	Dark green	1006, 2075, 2082
Food	Yellow	1031, 1058, 1065, 1070, 1104, 1106, 1110, 1112, 1116, 1118, 1308, 1348, 2002, 2008, 2012, 2014, 2016, 2018, 2022, 2026, 2028, 2030, 2032, 2124

shop very easily. Spreading out similar stores may encourage the customers to do more walking. In this way, every customer is exposed to every type of store. These factors may generate more total purchasing. Most malls use a combination of these principles, as you will see when you look at the example that follows.

The plan on page 17 is for the Arden Fair shopping center in Sacramento, California. It is the largest mall in the metropolitan area, which has a population of about 1.7 million people. Use a set of felt pens or colored pencils to color the various categories of stores as indicated in the table. After you have finished color-coding your plan, answer the questions that follow.

3. What patterns can you detect in the plan? Look in particular at "women's apparel" and "beauty/fitness and health" with respect to the two more prestigious anchor department stores (Nordstrom and Macy's).

_____

_____

_____

4. What shops might men be more interested in?

_____

_____

5. How are the jewelry stores, shoe stores, and places selling food grouped?

_____

_____

_____

6. For each of these types of stores, what can you say about their location within the mall?

_____

_____

_____

The Changing Geography of Ice Hockey

Ice hockey originated in northern Europe. In the 1870s, British soldiers stationed in Halifax, Nova Scotia, played on frozen ponds, and students at McGill University in Montreal played at a rink. Soon, amateur teams and leagues were organized in various Canadian cities. Toward the end of World War I, the National Hockey League (NHL) was established.

In this activity you will map the spread of the NHL, from its beginning in 1917 to 2001. In 1917, only four metropolitan areas had teams. In 2001 there were 30 teams. Follow the steps below to complete the map on page 21. Use it and data from the table to answer some questions about the changing geography of the NHL.

YOU ARE THE GEOGRAPHER

1. Use the list of metropolitan areas in the table on page 20 and an atlas to label each circle on the map with its correct place-name.

2. Use a set of felt pens to color each circle according to the time period when that metro area first obtained an NHL team. Use black for places that got a team in the period 1917–1920, brown for 1924–1935, green for 1967–1975, red for 1976–1981, and yellow for 1992–2000.

3. On a separate piece of paper, explain the pattern you see on the map. There are various factors you should think about. These include distance, climate, and population size. Examine your map and see if you think the following statement is a fair generalization: places that are near Montreal, that have cold climates, and that have relatively large populations were likely to get teams early, compared with places that are far from Montreal, have warm climates, and/or have relatively small populations.

   What happens to the pattern as we move forward in time? Do any of the three factors mentioned become less important? Why might that be the case? To help you answer these questions, the table includes latitude and longitude for each place and its population and population rank at the approximate time it first acquired a team.

   Canadians consider ice hockey to be their national sport and part of their national identity. Not only did Canadians start the NHL, but they continue to produce a very large number of its players. Canada exports hockey players not only to the United States but to other places where hockey is growing in popularity, including Europe and Japan.

4. How might you feel about the changing geography of ice hockey if you were a Canadian? Would you mourn the loss of NHL teams and players or celebrate the diffusion of your sport to the rest of the world?

Metropolitan Areas That Have (or Once Had) NHL Teams

Year acquired	Metropolitan Area	Latitude & Longitude	Population at the time	Rank at that time	Current Team(s)
1917	Montreal	45.30N, 73.35W	1,086,000	11	Canadiens
1917	Toronto	43.40N, 79.23W	901,000	14	Maple Leafs
1917	Quebec	46.49N, 71.13W	207,000	54	None
1917	Ottawa	45.25N, 75.43W	197,000	58	Senators
1920	Hamilton	43.15N, 79.52W	190,000	62	None
1924	Boston	42.15N, 71.07W	2,308,000	5	Bruins
1925	Pittsburgh	40.26N, 80.01W	1,954,000	7	Penguins
1925	New York-Northern NJ	40.40N, 73.58W	10,901,000	1	Rangers, Islanders, Devils
1926	Chicago	41.49N, 87.37W	4,365,000	2	Blackhawks
1926	Detroit	42.22N, 83.10W	2,105,000	6	Red Wings
1930	Philadelphia	40.00N, 75.13W	2,847,000	3	Flyers
1934	St. Louis	38.39N, 90.15W	1,294,000	8	Blues
1967	LA-Anaheim-Riverside	34.03N, 118.14W	7,176,000	2	Kings, Mighty Ducks
1967	SF-Oakland-San Jose	37.48N, 122.16W	4,399,000	6	Sharks
1967	Mpls.-St. Paul	44.58N, 93.15W	2,037,000	16	Minnesota Wild
1970	Buffalo	42.54N, 78.51W	1,590,000	29	Sabres
1970	Vancouver	49.76N, 123.06W	1,268,000	31	Canucks
1972	Atlanta	33.45N, 84.23W	1,832,000	19	Thrashers
1974	Wash., D.C.	38.50N, 77.00W	2,815,000	9	Capitals
1974	Kansas City	39.05N, 94.35W	1,293,000	30	None
1976	Cleveland	41.30N, 81.42W	1,950,000	18	None
1976	Denver	39.44N, 104.59W	1,464,000	23	Avalanche
1979	Hartford	41.45N, 72.40W	1,052,000	38	None
1979	Edmonton	53.33N, 113.28W	741,000	54	Oilers
1979	Winnipeg	49.53N, 97.09W	592,000	68	None
1980	Calgary	51.03N, 114.05W	626,000	61	Flames
1992	Tampa-St. Petersburg	27.57N, 82.25W	2,068,000	23	Lightning
1993	Miami-Ft. Laud.	25.45N, 80.11W	3,193,000	12	Florida Panthers
1993	Dallas	32.45N, 96.48W	3,885,000	9	Stars
1996	Phoenix	33.30N, 112.00W	2,111,000	22	Coyotes
1997	Raleigh-Durham	35.45N, 78.39W	735,000	61	Carolina Hurricanes
1998	Nashville	36.10N, 86.48W	985,000	42	Predators
2000	Columbus	40.00N, 83.00W	1,377,000	32	Blue Jackets

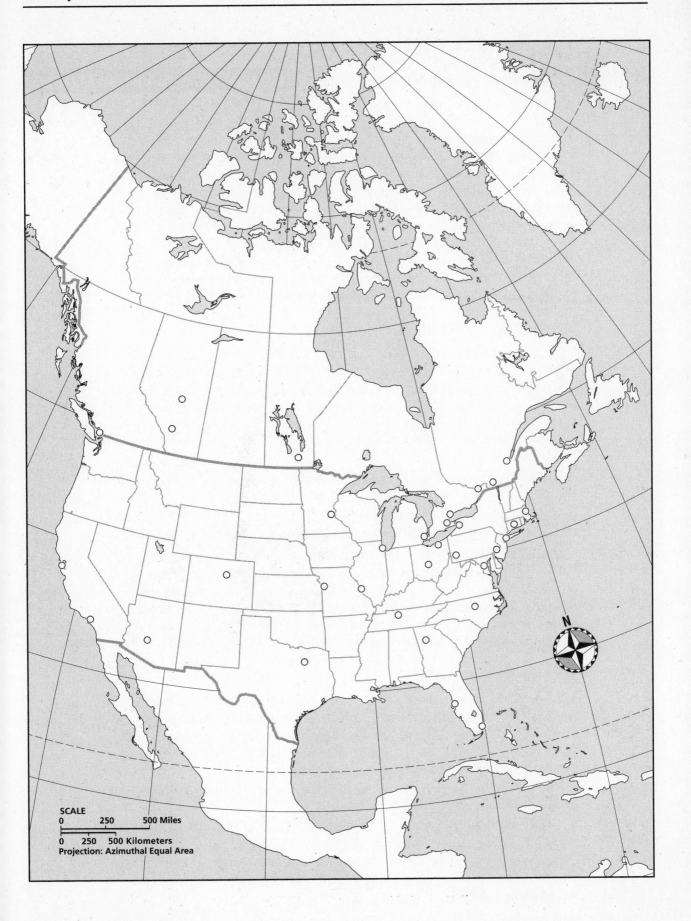

SCALE

0   250   500 Miles

0   250   500 Kilometers
Projection: Azimuthal Equal Area

Female Pioneers

The women's rights movement has influenced many aspects of life, including scholarship. It has raised awareness of the extent to which women's experiences and voices were ignored in many historical and geographical studies. For example, the subject of pioneering, or settling new lands, has been much studied by geographers. However, only recently has much attention been paid to how the pioneering experience might be different for women than for men. This exercise discusses what a team of researchers, including a geographer, learned about female pioneers in Mexico. You will write about the differences and similarities between their experiences and those of pioneer women in North America.

WHERE DID THE LAND GO?

Throughout most parts of Latin America, the distribution of land is very unequal. In fact, Latin America has the most unequal land distribution of any large region in the world. A few very wealthy people own a great deal of land. A great many poor people own little or none. Perhaps half of all agricultural lands belong to less than 2 percent of the landowners. Some 40 percent of all farm families have no land at all. This pattern has persisted from colonial times.

When Central and South America were first colonized, the land was simply taken away from the native peoples. It was then given to a few wealthy and influential Europeans. Today, it is clearly acknowledged that this practice was unfair. But what can be done to correct the problem?

While nearly all Latin American countries have said that they are using land redistribution programs, only a few have made much difference. Mexico, Bolivia, and Cuba are the three Latin American countries where revolutions resulted in some change in the established power structures. The failure of land reform can be attributed to the continuing power of wealthy landowning families and companies in most parts of Latin America.

SECOND-RATE LAND?

In Mexico, the government has tried to reduce the pain of redistributing existing farmland. The government has encouraged the landless to settle undeveloped lands. However, undeveloped does not mean vacant. These lands often had native occupants. In Mexico, these undeveloped lands are located in lowland, tropical parts of the country. These lands had not been settled before because of environmental difficulties including diseases such as malaria and yellow fever.

The communities where female pioneers were studied in 1990–1991 are located mostly in the interiors of Veracruz, Oaxaca, Tabasco, Chiapas, and Campeche. This region lies to the south of Mexico's central plateau, the economic and cultural core of the country. Find a map of Mexico and locate these states.

PIONEERING COMMUNITIES

Some pioneering communities in tropical Mexico consist of private farms. Others are *ejidos*. Under the *ejido* system, land is owned collectively by a village. Some plots are assigned to individual households. The right to work these private plots can be passed on to a widow or an offspring. Some settlements were planned and their settlers chosen. Other settlements were spontaneous—people just moved in.

The university researchers talked to the women of these pioneer communities for more than one year. They noticed some interesting patterns. Most of the women and their f— ew lands because of pove— ir hometowns. However, — difficult. There were no r— medical services, and there —

FROM BAD TO V—

While the male pi— n poorer than before. Perha— r breaking up of small agricul— of land only to household — always men. It was very h— ey in the small, isolated pion— not work in the fields. They — at brought in some income. Th— sewing and embroidery or — ly, the women used their s— household expenses.

The women felt very isolated. Many were used to living in large, extended families. In the new settlements, they had only their husbands and children. They spent most of their time at home doing housework and caring for children. Many felt that their household duties and their husbands' disapproval kept them from becoming friendly with neighbors. New neighbors often came from very different parts of the country. There were few public gathering places such as markets or parks where people could meet. Even the schools failed as gathering places for children and young people. Teachers would disappear for weeks at a time.

THE LAND DOES NOT PRODUCE

The poverty of many of these women was tied to the few ways they could make money. Even the land did not provide help. Tropical soils are fragile. After the forest had been cleared and crops had been raised for a few years, the soil lost its fertility. Productivity fell. This land was then switched to pasture for cattle.

Buying and caring for cattle is expensive. Securing a loan requires know-how that few poor farmers have. Only one fourth of the families in this study owned any cattle. To add to the problem, less labor is needed to tend to cattle and pastures than to raise crops. Thus, there is less work available. Now the men as well as the women were driven to look for additional ways to earn income.

Some studies of pioneering in the tropics have been very critical. This pioneering has resulted in the loss of tropical forests, the displacement of native peoples, and poverty for many pioneers. Governments have spent huge sums on pioneering projects. Perhaps the money would have been better spent improving people's lives where they were already living.

YOU ARE THE GEOGRAPHER

In a small group discussion, recall what you know about the North American pioneer experience. Perhaps you have read books on pioneering by Laura Ingalls Wilder, novels by Willa Cather, or excerpts from the journals and diaries of pioneer women. Think about the similarities and differences between the experiences of female pioneers in North America in the 1800s and those in Mexico in the 1900s. Topics might include encountering new natural environments, farm and household tasks, or dealing with isolation. Write down your thoughts in a short essay.

_____

_____

_____

_____

_____

_____

_____

_____

_____

_____

_____

_____

Planning for Tourism in Dominica

GEOGRAPHY FOR LIFE ACTIVITIES
NATIONAL GEOGRAPHY STANDARD 18

The travel and tourism industry has become the world's largest. It is of major economic importance to many countries and regions. It is of great interest to geographers because it involves so many of geography's key themes: movement, human-environment relationships, and the character of places.

Many tropical countries that are former European colonies used to depend on one or two export crops (such as sugar, bananas, or coffee) as the basis of their economies. Now they have become very dependent on another type of "crop"—tourists. This shift is true of many Caribbean countries. Small island countries that have fewer economic choices are the most dependent on tourism.

THE GOOD AND THE BAD OF TOURISM

Tourism has both positive and negative consequences. It can bring money and jobs and better roads, sewers, and parks. It can boost other local industries such as food production, construction, and manufacturing. These jobs, however, can be seasonal and low paying. Money can flow back out of the country if the investors in the tourist industry are foreign firms. Money also leaves the country if many of the goods needed by tourists have to be brought in. Tourism can provide resources for a country to conserve its cultural monuments and traditions.

Tourism can also lead to overcommercialization of what was serene and sacred. Contact between visitors and hosts can result in cross-cultural understanding or it can produce resentment. Tourism can bring about greater environmental awareness and conservation, but it can also bring decreased environmental quality. Problems such as congestion, pollution, and habitat loss can increase. Focusing only on tourism is risky. The flow of tourists can be stopped by economic downturns, political instability, or natural disasters.

SPECIALTY TOURISM

The tourism industry has become increasingly segmented in recent years. There are different types of tourists and different types of tourism. Not everyone wants to join crowds lying on the beach, spending money in duty-free shops, or gambling in casinos. Special interest tourism involves smaller groups of people interested in a particular focus or activity during their trip.

The focus of special interest might be fishing, diving, boating, or learning about the local culture, architecture, or natural history. A number of Caribbean islands are trying to avoid the problems of mass tourism. They have decided to focus on small-scale, high-quality, special interest tourism.

DOMINICA'S NEW FOCUS

One such island country is Dominica. Tourism did not develop early on Dominica because of its rugged topography (86 percent of the island has slopes of more than 20 degrees; the only flat land is a narrow coastal strip). The lack of white sand beaches and the distance from the United States also slowed tourism. On the steep slopes of its volcanic mountains stand some of the Caribbean region's last remaining old growth tropical rain forests. In fact, forests still cover 60 percent of Dominica. Eleven different plant communities and diverse wildlife, including numerous endemic (only occurring locally) species, have been found on Dominica. The most famous examples are two brightly colored parrots, the Imperial and the Red-Necked.

The island has numerous other natural attractions, including streams, pools, and waterfalls. The west and north shores have the best beaches. There is also an underwater world of coral reefs off Dominica's shores. Because of these unparalleled natural beauties, Dominica has chosen to present itself as the "Nature Island of the Caribbean." Of cultural interest are the Carib Indians (one of only two surviving populations), historical monuments, and Dominica's mix of African, Carib, French, and English cultural elements.

Tourism services currently available include small hotels and guest houses, restaurants, discos, tour operators, and rental car agencies. Most of these businesses are found on the drier west coast of the island. There are two airports, although neither can handle large jets. There are two ports where cruise ships can berth. Many places on the island have no paved roads, and most roads are threatened by frequent landslides. Even the most popular natural attractions lack basic facilities such as trails, toilets, shelters, and signs.

YOU ARE THE GEOGRAPHER

Use the information in this exercise, the table on page 27, and a map of Dominica, plus any additional resources to discuss and answer the following questions about Dominica's efforts to develop its nature tourism potential. Write your answers on a separate sheet of paper.

1. Who should be making the decisions about tourism development? How should the decision-making process work?

2. What kinds of information would be useful for those planning tourism development on Dominica? List as many kinds as you can.

3. The government has been trying for years to develop a larger airport for big jets. Do you think this is a good idea? Why or why not? Where on the island would you put it (remember the only level land is along the coasts, and even there it is very limited)?

4. The government spent millions of dollars to build a second cruise ship facility south of Cabrits National Park. Is cruise ship tourism an approach that Dominica should be pursuing? Does this mix well with the island's desire to present itself as "the nature isle of the Caribbean?" If cruise ships come, how can the island maximize income from them, yet minimize the impacts from the sudden arrival of hundreds of people all at once?

5. If more overnight accommodations are needed, what types should they be (large hotels, hotel resorts, small hotels, bed and breakfast inns, rooms in people's homes, camping facilities, etc.)? What should they look like? Where should they be located?

6. What connections could be made between tourism development and investment in other economic sectors on the island (agriculture, construction, manufacturing)? The idea is to have the tourism sector buy whatever it needs locally, to keep money on Dominica.

7. What kinds of things need to be done to protect the environment as more tourists come to visit? How would these programs be paid for?

Selected Characteristics of the Commonwealth of Dominica	
Area	300 sq. mi. (about 27 mi. by 11 mi.)
Topography	Rugged and steep; highest elevation: 4,747 feet
Climate	Humid tropical marine (average annual temperature on coast—80°, on highest peaks—70°; average annual rainfall, 50″–400″; prevailing winds are from the northeast)
Population	70,000
Principal languages	English (official), French dialect
Per capita gross domestic product	$4,000
Government	Parliamentary democracy
Capital city and major port	Roseau (population: 16,000)
Industries	Tourism, coconut products, furniture, cement blocks, shoes
Major crops	Bananas, citrus, mangoes, coconuts
Natural resources	Forests, abundant freshwater

Regional Images of the Venezuelan Andes

A region is an area of Earth's surface that has characteristics that make it different from surrounding areas. Regions vary greatly in size. A neighborhood, a continent, or a hemisphere can be a region. An area becomes a region when someone recognizes that it is different in some respect from adjacent areas. There are many ways in which one area differs from another, and thus there are many possible types of regions.

HOW REGIONS ARE DETERMINED

Sometimes physical features are used to divide Earth or some part of it into regions. You may have seen maps of landform regions, vegetation regions, or climate regions. Sometimes cultural features are used. You may have seen maps that divide areas according to language or religion. For example, maps of Canada often show its French-speaking and its English-speaking regions.

Economic data can also be used to draw regional boundaries. For example, if you were a store manager, it might be very useful to know from what territory you draw customers. If you mapped their addresses and drew a boundary around them, you would have a map of the region served by your store. This information might influence your decisions about where to advertise.

Many regions are recognized on the basis of more than one characteristic. For example, Appalachia is a well-known region of the United States. People associate with Appalachia certain physical features, natural resources, ways of making a living, and cultural and social features. These different characteristics are interrelated. For example, the presence of coal made possible the development of coal mining as a major economic activity in the region. Geographers are particularly interested in how natural features and human history have come together to give a region its distinctive character.

REGIONS CAN CHANGE

Regions and how they are perceived change through time. This is because things change within a region. For example, a major coal seam is mined out and the mines close down. The region is no longer a major coal producer, nor is it perceived as such. The boundaries of regions can also change as people, economic activities, transportation links, administrative centers, and even physical features shift their locations.

REGIONS IN THE ANDES

Geographer Marie Price has written about regional change through time in the Andes Mountains of Venezuela. (Use an atlas to see how the northeastern spur of the Andes, the Cordillera de Merida, cuts across northwestern

Venezuela south of Lake Maracaibo.) She has suggested that since the mid-1800s the regional image of the Venezuelan Andes had passed through three phases: regional integration, regional decline, and cultural refuge.

Regional Integration   In the regional integration period from the 1870s to the 1930s, the dominant economic activity was raising and exporting coffee. Coffee growing spread through the *tierra templada* (the temperate zone between 3,000 and 6,000 feet elevation). Exporting coffee was made easier by the construction of transportation links. The Trans-Andean Highway ran the length of the region, tying the coffee towns together and providing the region with better access to Venezuela's capital city, Caracas. Several rail lines led to the shipping lanes on Lake Maracaibo. The coffee reached the port city of Maracaibo.

The population of the region grew faster than that of Venezuela as a whole. Population density was higher. The region had large estates and small independent farms. The traits of hard work, independence, and conservativism became part of the image of the *andinos* (residents of the Andes). Interestingly, such traits are associated with mountain people in many parts of the world. They were celebrated by Andean cultural societies, schools, and newspapers.

Regional Decline   In the regional decline period, from the 1940s to the 1960s, the Venezuelan Andes fell on hard times. Coffee prices collapsed in the 1930s as the Great Depression gripped the world. This collapse resulted in numerous small farmers losing their land and becoming tenants. Agricultural productivity fell. Oil was discovered in the lowlands (around Lake Maracaibo and elsewhere). It replaced coffee as Venezuela's most important export.

Increasingly, investments were directed to the oil regions and not to the Andes. New roads were constructed, but they bypassed the region. The Pan-American Highway of 1955 was built along the northern base of the mountains. The Llanos Highway of 1969 was built along the southern base. Along the new roads, which were in the *tierra caliente*, migrants from the highlands planted bananas and tobacco or raised cattle.

Andinos began to leave the region at much higher rates than before. By the end of this period, there were 1.2 million people living in the region, but 1.8 million *andinos* had left and lived elsewhere in Venezuela. The dominant traits of *andinos*, previously admired, were now criticized. Peasant farmers were accused of being old-fashioned and unwilling to change. The Venezuelan Andes had become a backwater.

Cultural Refuge   In the cultural refuge phase, from the 1970s to the present, the economic base and the dominant image of the Venezuelan Andes changed again. Agriculture continued to be important in places. New types of agriculture appeared, including the raising of vegetables and flowers in the *tierra fria* (6,000 to 9,000 feet). As a whole, the population and the economy have become urban. More than two thirds of the Andean population live in cities and work in service or manufacturing jobs. Tourism, both domestic and international, has become an important new source of income for the region.

As Venezuelans have become increasingly modern, urban, and mobile, they have come to treasure the very qualities of the Andes region that previously were criticized. Traditional ways of doing things are no longer viewed as ignorant and backward. Now these ways are quaint, unique, and worth preserving. Traditional villages and buildings have been restored or recreated. Artists produce images of the region that celebrate its natural beauty and rural past.

The region remains somewhat isolated from the rest of the country, both economically and culturally. However, it has become more central to the identity of Venezuela as a place where nostalgia can be indulged without holding back the general progress of the country.

An important part of Marie Price's work on the Venezuelan Andes is her argument that regions are not just what is out there on the ground. What is out there is important, but human perception and imagination interact with the environment to form regions. It is important to remember that regions are human constructs, and that these constructs are always changing.

YOU ARE THE GEOGRAPHER

Make a table that summarizes the information presented above. Across the top, label three columns with the dates of the three periods of time used in the discussion. Down the left side, put categories of information such as "people," "economy," "transportation," and so on. Fill out the table. Next, pick a different region for which you can make a similar table. It might be the region where you live (such as your state) or a region that you have previously studied in school. Make a table like the one you made for the Venezuelan Andes. See if you can come up with several distinct time periods in the life of the region and then fill in the traits that characterized the region in each of those time periods. What sources influence your ideas about this region? Schoolbooks? Other books? Maps? Newspapers? TV and movies? Works of art? People you have talked to? Your personal observations?

ACTIVITY 11

Mapping Wealth and Poverty in Brazil

GEOGRAPHY FOR LIFE ACTIVITIES

NATIONAL GEOGRAPHY STANDARD 1

There are many different types of maps. The type you will construct in this exercise is a choropleth map. Choropleth maps are used to display ratio data (bushels per acre, TVs per household, income per capita, etc.). You will be using colors to symbolize different values of gross domestic product (GDP) per capita (a measure of income) for Brazil's 26 states and Federal District.

The table on page 32 displays the 1997 GDP data just as they are provided by the Brazilian Institute of Geography and Statistics. The figures are given in Brazilian *reals* (R$). In 1997 R$1.09 was equal to U.S.$1.00. GDP per capita is given for the country as a whole, for Brazil's five regions (Norte, Nordeste, Sudeste, Sul, and Centro-Oeste), and for the 26 states and the Federal District within the five regions. Population figures are included for your information but are not needed in making the map of income per capita by state.

YOU ARE THE GEOGRAPHER

Follow the steps below to construct your map. Use the map on page 33.

1. List the GDP figures for the states and the Federal District in order in a column, starting with the lowest and ending with the highest. Now you must divide the data into classes. Usually no more than eight classes are used on a choropleth map, because too many classes make the map too difficult to read. For this map, use 4, 5, or 6 classes. Divide the data into groups with equal (or as nearly equal as possible) numbers of states. For example, if you decided to use five classes with your Brazil income data, each class would have 5 or 6 states in it.

2. Pick a color to represent each class. Avoid using colors or patterns that are all the same value (degree of darkness or lightness). A scheme such as light blue, medium blue, dark blue, purple, black is better than a scheme such as medium blue, medium green, medium red, medium orange, medium brown. This is because map readers make an assumption that lighter colors mean lower values and darker ones mean higher values.

3. Color your map according to your scheme. Do not forget to make a key, give your map a title, and cite the source of information. What can you say about the GDP in Brazil and the physical characteristics, resources, soil, and climate of the various regions?

Brazil has the widest income gap in the world: the top 10 percent of the population gets almost 50 percent of the national income. The bottom 50 percent gets little more than 10 percent. The richest Brazilian state, São Paulo, has a per capita GDP more than six times that of the poorest state, Maranhão. By contrast, in the United States, the richest state has a per capita income only double that of the poorest state.

BRAZIL'S GROSS DOMESTIC PRODUCT PER CAPITA BY STATE AND REGION, 1997

Area	Population (in millions)	GDP per capita (in R$)
BRAZIL	157.1	5,413
NORTE	11.3	3,293
Rondonia	1.2	3,317
Acre	0.5	2,605
Amazonas	2.4	5,816
Roraima	0.2	2,423
Pará	5.5	2,584
Amapá	0.4	3,767
Tocatins	1.0	1,580
NORDESTE	44.8	2,494
Maranhão	5.2	1,389
Piauí	2.7	1,555
Ceará	6.8	2,522
Rio Grande do Norte	2.6	2,551
Paraîba	3.3	2,082
Pemambuco	7.4	3,115
Alagoas	2.6	2,145
Sergipe	1.6	2,900
Bahia	12.5	2,890
SUDESTE	67.0	7,436
Minas Gerais	16.7	5,118
Espírito Santo	2.8	5,639
Rio de Janeiro	13.4	7,152
São Paulo	34.1	8,822
SUL	23.5	6,402
Paraná	9.0	5,736
Santa Catarina	4.9	6,380
Rio Grande do Sul	9.6	7,036
CENTRO-OESTE	10.5	5,008
Mato Grosso do Sul	1.9	4,693
Mato Grosso	2.2	3,972
Goiás	4.5	3,428
Distrito Federal	1.8	10,508

ACTIVITY 12

Agricultural Patterns of Western South America

GEOGRAPHY FOR LIFE ACTIVITIES

NATIONAL GEOGRAPHY STANDARD 15

Western South America (Ecuador, Peru, Bolivia, and Chile) is a diverse region agriculturally. It includes both subsistence and commercial types of farming. Subsistence farmers, who grow only enough to feed themselves, are concentrated in the Andean parts of Ecuador, Peru, and Bolivia. The staple foods that they produce include barley, maize, and potatoes. Grazing is also important. In addition to these long-established farmers of the Andes, Western South America is also home to some shifting agriculture (slash and burn) in the tropical portions of the region.

YOU ARE THE GEOGRAPHER

This activity focuses on commercial agriculture. It looks closely at those farm products that are exported. Use the data on page 35, together with maps from your text or an atlas, to answer the questions. In those questions that ask you to rank the countries, always start with the country with the highest value and work down to the lowest value. Read through the entire exercise before you begin answering the questions.

1. Rank the four countries in order of total merchandise exports.

_____

_____

2. Rank the four countries in order of total agricultural exports.

_____

_____

3. Calculate for each country what share of its total merchandise exports is accounted for by agricultural exports. Write the countries and figures in rank order.

_____

_____

4. Calculate for each country how much it receives per capita from its agricultural exports. Write the countries and figures in rank order.

_____

_____

5. Which country's agricultural imports have a higher value than its exports? Which country's agricultural exports equal about four times its agricultural imports?

_____

_____

Table 1. Selected Statistics for Bolivia, Chile, Ecuador, Peru, and South America

	Total Merchandise Exports (U.S.$1,000)	Total Agricul- tural Exports (U.S.$1,000)	Total Agricul- tural Imports (U.S.$1,000)	Total Popu- lation	Irrigated Land (Hectares)
Bolivia	1,151,700	417,600	163,100	7,957,000	88,000
Chile	17,024,800	2,540,800	1,292,200	14,824,000	1,270,000
Ecuador	5,264,400	1,943,700	486,900	12,175,000	25,000
Peru	6,754,000	732,200	1,309,500	24,797,000	1,760,000
South America	148,149,900	41,113,900	16,039,600	464,266,800	9,902,000

Table 2. Selected Export Statistics for Bolivia, Chile, Ecuador, Peru, and South America
Each figure is given in thousands of U.S. dollars (U.S.$1,000).

	Apples	Bananas	Cocoa Beans	Coffee Beans	Cotton Lint	Grapes
Bolivia	0	254	369	26,050	39,254	0
Chile	189,581	18	0	10	0	413,954
Ecuador	0	1,326,977	59,185	86,300	4,517	0
Peru	59	6	18	357,620	31,683	1,874
South America	331,910	1,869,504	81,141	5,498,790	484,150	438,641

Table 2, continued.

	Maize	Milk	Peaches	Pears	Pineapples	Rice
Bolivia	630	792	0	0	65	2
Chile	50,240	26,042	26	73,749	17	5
Ecuador	31,120	1,044	0	0	2,599	46,402
Peru	2,350	1,138	57,316	0	1	3,803
South America	1,501,850	324,834	61,769	251,490	7,193	671,610

Table 2, continued.

	Rubber, Natural	Soybeans	Sugar, Raw	Tomatoes	Wine
Bolivia	0	228,766	22,050	1	75
Chile	81	1,351	70	1,716	427,931
Ecuador	668	528	2,400	0	8
Peru	0	0	34,290	8	213
South America	953	9,642,106	2,269,990	4,860	576,864

6. Calculate what share of South America's total agricultural exports come from these four countries.

_____

_____

7. In which export crops does Ecuador lead this group of four countries? What kind of climate suits this group of crops? Now look at a climate map of South America. Peru and Bolivia also have areas with this climate. What is different about where this climate is found in Ecuador, and why is it a significant locational factor?

_____

_____

8. In which export crops does Peru lead this group of four countries? Which one of these do you think is a _tierra templada_ crop? Sugar is grown on the coastal plain in what climate type (Peru's entire coast is in the same climate zone)? How can that occur? (Look at the column titled "Irrigated Land.")

_____

_____

9. In which export crops does Bolivia lead this group of four countries? The larger of these is now Bolivia's biggest export. Where do you think these are grown in Bolivia? Hint: think about where they are grown in the United States and that will help you figure out where they are grown in Bolivia.

_____

_____

10. In which export crops does Chile lead this group of four countries? Match up these crops with the climate regions in Chile where you would be likely to find them. Chile's mix of agricultural products is like which U.S. state's? For which of these crops does Chile's output represent more than 50 percent of the total for South America?

_____

_____

The questions above help to illustrate the relationship between climate (which in turn is related to location) and agriculture. What gets produced in a place is also determined by many other factors. Other important aspects of the physical environment include soils and slope.

In the case of commercial agriculture, there needs to be a market for the crop. Money must be available to invest in land, water, seed, fertilizer, and machinery. Labor can be a very important consideration, depending on the crop. Expertise is also needed. Storage and transportation have to be available to deliver the product to market in good shape.

The Mediterranean Sea

GEOGRAPHY FOR LIFE ACTIVITIES

NATIONAL GEOGRAPHY STANDARD 14

The Mediterranean Sea has shaped the civilizations on its shores for thousands of years. This activity explores the geography of the Mediterranean Sea and the ways in which people have used its resources. Read through the entire activity once. Then use an atlas to locate and label all the places underlined below on the outline map on page 39. Use dots for the cities and draw in the rivers.

The word *Mediterranean* means "in the middle of land" or "inland" in Latin. Europe is to its north, Asia to its east, and Africa to its south. Twenty-one countries share the Mediterranean today. In Europe these are <u>Spain</u>, <u>France</u>, <u>Monaco</u>, <u>Italy</u>, <u>Malta</u>, <u>Slovenia</u>, <u>Croatia</u>, <u>Bosnia and Herzegovina</u>, <u>Yugoslavia</u>, <u>Albania</u>, and <u>Greece</u>. In Asia they are <u>Turkey</u>, <u>Cyprus</u>, <u>Syria</u>, <u>Lebanon</u>, and <u>Israel</u>. In Africa they are <u>Egypt</u>, <u>Libya</u>, <u>Tunisia</u>, <u>Algeria</u>, and <u>Morocco</u>.

GEOGRAPHY

Covering about 1 million square miles and stretching over 2,300 miles in length, the Mediterranean is a large body of water. Only the world's oceans and the Arabian Sea are bigger. The Mediterranean is divided into two basins—western and eastern—by a relatively shallow area of the sea between Tunisia and Sicily.

These basins are further subdivided into a number of smaller seas: the <u>Ligurian</u> and the <u>Tyrrhenian</u> Seas in the western basin and the <u>Adriatic</u>, the <u>Ionian</u>, and the <u>Aegean</u> in the eastern basin.

Thousands of islands dot the Mediterranean Sea. The largest islands are <u>Sicily</u>, <u>Sardinia</u>, <u>Cyprus</u>, <u>Corsica</u>, and <u>Crete</u>. Greece has the most islands. Spain's <u>Balearic Islands</u> are another famous group.

Several straits have long histories of strategic importance. These include the <u>Strait of Gibraltar</u>, linking the Mediterranean Sea with the <u>Atlantic Ocean</u>, and the <u>Dardenelles</u> and <u>Bosporus</u> (with the <u>Sea of Marmara</u> in between), linking the Mediterranean with the <u>Black Sea</u>. The <u>Strait of Messina</u> and <u>Strait of Otranto</u> are other narrows in the Mediterranean that have been used and fought over for millennia.

CLIMATE

The influence of the Mediterranean on the peoples in its vicinity has been enormous. The sea is a moderating influence on climate. It cools the shores during the region's hot summers. The coastal winters are mild in contrast to the colder winter conditions in the nearby hills and mountains. The sea has long supplied the nearby human populations with food and other goods. However, its greatest significance lies in its function as a means of transportation and communication.

Several characteristics of the Mediterranean Sea make it particularly useful for navigation. Its tides are minor. Its waves are generally smaller than those that develop on open oceans and seas. The sea's many peninsulas and islands mean that ships are never far from land. The coastline and narrow continental shelf provide many excellent natural harbors, particularly on the northern (European) coast of the sea.

TRADE

In ancient times, the great civilizations of the Mediterranean region navigated the sea in order to trade and to establish new colonies. Places well beyond the Mediterranean also sent their goods to the region to be traded. In this way, various civilizations came into contact with one another, leading to new cultural connections and innovations. In later centuries, the Mediterranean declined somewhat in commercial importance. When the Suez Canal, connecting the Red Sea with the Mediterranean, opened in 1869, it greatly enhanced the Mediterranean's role in the global transportation network.

ENVIRONMENTAL PROBLEMS

Today the Mediterranean Sea is suffering from a number of problems. It is one of the most oil-polluted seas in the world. This pollution comes from a variety of sources, including tankers, offshore oil rigs, oil refineries and petrochemical plants, and discharge from sewers and rivers. Industrial pollutants, including heavy metals and synthetic compounds, enter the sea in many places, but are especially concentrated near the mouths of the Rhone and the Po Rivers. Much industrial activity is found in the major port cities. In Europe, these include Algeciras (on the Bay of Gibraltar), Valencia, Barcelona, Marseilles, Genoa, Venice, and Piraeus (the port city for Athens).

Runoff from the Mediterranean region's orchards, vineyards, and fields contains herbicides, pesticides, and fertilizers. Fertilizers contribute to algae blooms, which die and rot, using up the sea's oxygen. The lack of oxygen kills organisms in the water and on the seabed. The discharge and breakdown of human and animal waste have the same effect.

The Mediterranean region is also a magnet for tourists. More than 150 million visitors arrived in Spain, France, and Italy in 1998. Many of these tourists are attracted to the sun, sea, and sand of the Mediterranean. Examples of heavily developed tourist zones are the Costa Brava (North of Barcelona), the Costa Blanca (south of Valencia), and the Costa del Sol (around Malaga) in Spain and the Riviera in France, Monaco, and Italy (from Cannes to La Spezia). All those extra people put more stress on the water, air, and land of the region. The problems of pollution, inadequate water supplies, and loss of habitat and open space are worsened.

The environmental problems of the Mediterranean are being addressed by individual countries and by international organizations, including the European Union and the United Nations. Because the Mediterranean is shared by 21 different countries, international cooperation is very important for success.

Mapping the Mediterranean

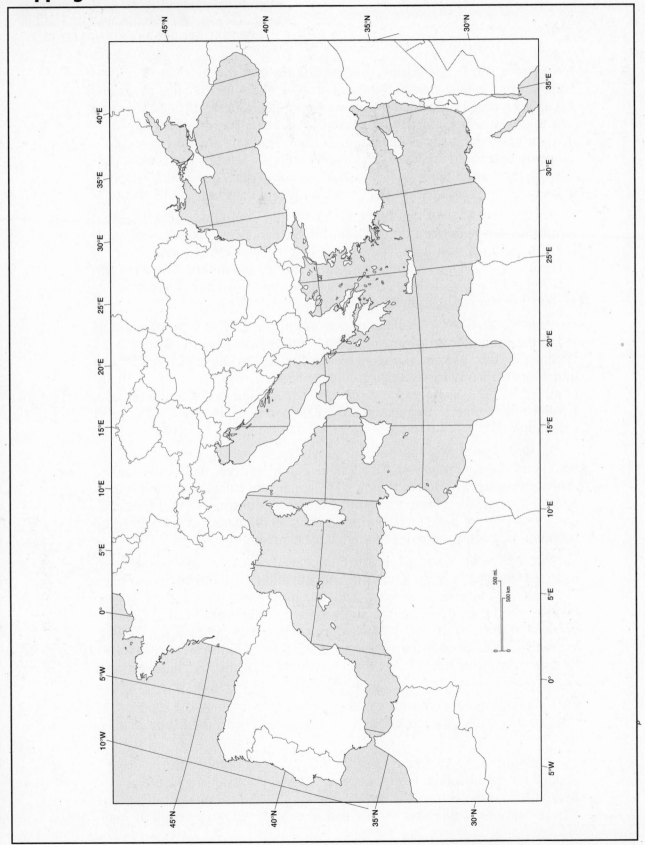

The EU and NATO: Past, Present, and Future

GEOGRAPHY FOR LIFE ACTIVITIES

NATIONAL GEOGRAPHY STANDARD 18

After World War II, the economy of every European country was in ruins. Some European leaders realized that Europe had a better chance of restoring its economic health if countries worked together and opened their markets to one another rather than remaining isolated. These leaders hoped that with free trade and cooperation, Europe would be able to create an economy that could compete successfully with the world's two new superpowers, the United States and the Soviet Union. Furthermore, this economic cooperation would reduce the chance of future conflicts within Europe.

On January 1, 1958, the European Economic Community (EEC) came into being. It was formed by Belgium, the Netherlands, Luxembourg, Germany, France, and Italy. The coming together of Germany and France was particularly significant after several centuries of conflict. The countries that formed the EEC dedicated themselves to the creation of a common market and the removal of trade barriers.

At first, Great Britain decided to stay out of the EEC. It was still committed to the economic arrangements it had with members of the British Commonwealth. Britain attempted to join the EEC in the 1960s, but its entry was vetoed twice by France, a long-standing rival. Britain was finally admitted in 1973, along with two countries that have strong trade ties to it, Ireland and Denmark. By this time, the EEC had become the EC (European Community).

In 1981 Greece was approved as a member of the EC, and in 1986 Spain and Portugal followed. These three countries were poorer than the others (together with Ireland, they were sometimes called "the poor four"). Special arrangements were made for these new countries, so that they could join but not have their own economies immediately overwhelmed by the more efficient economies of the older EC members.

In 1993 the EC became the EU (European Union). This new name reflected the greater ambitions of the organization. The EU wanted to have a completely integrated economy. This integration included the free movement of goods, services, capital, and people and the creation of a common currency (the euro). It also meant more agreement on political, social, and even defense matters. Some people used the term "United States of Europe" to suggest how closely tied the countries of the EU might become.

In 1995 Austria, Finland, and Sweden joined the EU. All of these countries are prosperous democracies. Even before joining, they did much of their trading with EU countries and worked closely with the EU on many issues.

Today there are 15 members of the EU. Thirteen other countries have applied for membership: Malta, Cyprus, Turkey, Bulgaria, the Czech Republic, Estonia, Hungary, Latvia, Lithuania, Poland, Romania, the Slovak Republic, and Slovenia. The last 10 of these countries recently had communist governments and centrally planned economies. They were satellite states of the Soviet Union until its collapse in the 1990s. The EU is helping these countries to create democratic and capitalist systems so that they can become full-fledged members.

A second important international organization with a European focus is the North Atlantic Treaty Organization (NATO). This military alliance was created in 1949 by a group of countries committed to each other's defense: Belgium, Canada, Denmark, France, Iceland, Italy, Luxembourg, Netherlands, Norway, Portugal, the United Kingdom, and the United States. In 1952 Greece and Turkey joined; in 1955, Germany; in 1982, Spain. The most remarkable expansion of NATO occurred in 1999, when former Soviet allies Poland, Hungary, and the Czech Republic joined NATO. With the Soviet Union gone and the Cold War over, NATO has turned some of its energies to peacekeeping efforts, a role it played in the recent Balkan troubles.

YOU ARE THE GEOGRAPHER

In this exercise, you will construct a map showing the growth of these important international organizations. Use the map on page 42 to show the growth of the EU and of NATO.

1. Use an atlas or your textbook to correctly label all the countries mentioned in the above discussion.

2. Use colored pencils or felt pens to color the EU countries as follows: the six original members (1957)—green; the next 3 (1973)—blue; Greece (1981)—purple; Spain and Portugal (1986)—orange; Austria, Sweden, and Finland (1995)—yellow. Color the 13 countries that have applied for membership in red stripes.

3. Use a black pen to indicate the NATO countries as follows: put horizontal stripes across the 12 original members; put vertical stripes down the 4 members that joined between 1952 and 1982; put diagonal stripes on the 3 that joined in 1999.

4. Put a title on your map and make a key for it.

5. Notice that Iceland and Norway are not members of or applicants to the EU. Notice that Sweden, Finland, Ireland, and Austria are not members of NATO. Switzerland, Albania, and most of the countries that used to be part of Yugoslavia are both non-EU and non-NATO countries. Have different people in your group find out why one of these countries has not joined these organizations. (For example, Iceland worries about other countries being able to fish in its waters if it joins the EU.) Other students could find out what are the challenges that the 13 EU applicants have to meet to become members. Share your findings.

The Vikings Abroad

GEOGRAPHY FOR LIFE ACTIVITIES

NATIONAL GEOGRAPHY STANDARD 9

Vikings is the term we use today for the people who, from the A.D. 700s to the 1000s, left their homelands in Scandinavia for other lands. They began three centuries of raiding, trading, exploration, and colonization. The Viking movements are a good example of migration.

REASONS FOR MIGRATION

Many people have wondered what prompted the Viking expansion. People migrate, or permanently relocate, in response to push and pull factors. Push factors are things about a person's current location that are undesirable. These factors may "push" the person to leave. Pull factors are the things about a new location that are attractive and "pull" a person toward it. Not enough land for farming, herding, or hunting was an important push factor for the Vikings. Also, some Vikings wanted to escape the rule of a particular king or chief.

Many historians now think that the increasing wealth of Europe was the biggest pull factor. By the 700s, trade was picking up again after having collapsed with the fall of the Roman Empire. Treasure was piling up in churches, monasteries, and other religious centers. Increased opportunities to plunder attracted the Vikings.

BARRIERS TO MIGRATION

For any group or individual, there are also barriers to migration. These can be physical, cultural, economic, or legal. The major barrier to migration for the Vikings was that the native peoples did not want them in their land. In some places, the Vikings were defeated militarily. In other places, they raided successfully. They gained treasure and slaves, but no land. In yet other places, the Vikings were able to establish towns and villages and bring in hundreds or thousands of settlers. Some people paid the Vikings to go away. These bribes were called Danegelds. Others offered the Vikings land to settle on; this is what happened in Normandy. Yet other groups invited the Vikings to become their rulers. These groups included the Slavic people in the area of present-day Novgorod (see map). In this way, the Vikings, who were sometimes called the Rus, became the founders of Russia. It is clear to see from the Viking example that immigrants may receive a variety of responses from others when they arrive.

EFFECTS OF MIGRATION

Migration can have many consequences. For example, the Vikings who settled in the lands where they were raiders or traders mixed into the local society. They learned the local language, married local people, and adopted new religions such as Christianity.

The Viking expansion had many effects on people's lives. Sometimes trade was disrupted, but at other times it was improved. Boat design improved with Viking technology. Numerous place-names in eastern England developed from Viking words. Words such as *window, husband, sky, anger, low, ugly, wrong, happy, thrive, ill, die, bread,* and *eggs* came into English this way.

The Vikings brought democratic ideas with them to Iceland. There they established the Althing, the world's first parliament. Upon arrival, they also immediately set to work transforming the landscape of the island, heavily exploiting the walrus (which apparently died out rather quickly) and the great auk (which eventually became extinct in the 1800s). Wood-cutting and the introduction of domesticated animals and plants led to the destruction of much of the natural vegetation of the island.

The effects of the Viking emigration on their homelands were several. Wealth from raiding and trading poured into the region. Some wealth was undoubtedly carried back to Scandinavia by returning migrants. Some Vikings went back to their old homes, rather than staying in the new land. For every migration stream, there is a return migration stream. People who returned were either disappointed with the new place or they had always intended to go home.

YOU ARE THE GEOGRAPHER

Use the map on page 45 and an atlas to answer the following questions.

1. What three modern countries were the Viking homelands?

_____

2. From which islands did Vikings go west to Iceland and beyond?

_____

3. From which countries did Vikings go south to western and southern Europe?

_____

4. From which country did Vikings go east to eastern Europe, Russia, and Asia?

_____

5. What is the earliest date on the map for an area "occupied or dominated" by Vikings? Where is it? This place is part of what country today?

_____

Now, write an essay that compares and contrasts a modern example of migration to the Viking example. This could be the migration story of an individual or family with whom you are familiar or of a large group such as Scandinavians coming to the United States after the Civil War or Mexicans drawn to California or Texas in the 1900s. Discuss push and pull factors, barriers to migration, and the consequences of the migration. Explain how your modern-day example of migration is similar to and different from the experience of the Vikings.

The Viking Migration

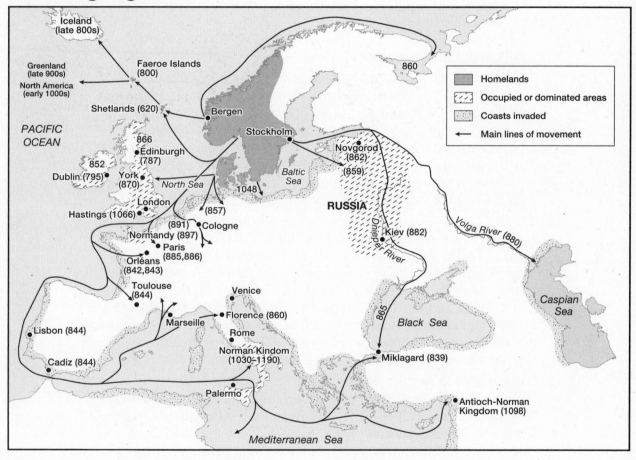

Iceland
(late 800s)

Greenland
(late 900s)

North America
(early 1000s)

Faeroe Islands
(800)

Shetlands (620)

Bergen

PACIFIC
OCEAN

866
Edinburgh
(787)

852

Dublin (795)

York
(870)

North Sea

London

Hastings (1066)

(891)

(857)

Cologne

Normandy (897)

Paris
(885,886)

Orleans
(842,843)

Toulouse
(844)

Venice

Florence (860)

Marseille

Rome

Lisbon (844)

Norman Kindom
(1030–1190)

Cadiz (844)

Palermo

Mediterranean Sea

Stockholm

Baltic
Sea

1048

Novgorod
(862)

(859)

RUSSIA

Kiev (882)

Dnieper River

Volga River (880)

865

Black Sea

Miklagard (839)

Antioch-Norman
Kingdom (1098)

Caspian
Sea

860

| Homelands |
| Occupied or dominated areas |
| Coasts invaded |
| → Main lines of movement |

Cities and Rivers in Eastern Europe

GEOGRAPHY FOR LIFE ACTIVITIES

NATIONAL GEOGRAPHY STANDARD 12

A city needs a good site and a good situation to grow in size and power. Good sites often have flat, well-drained land, available water, and (in the past) defensibility from attack. In almost all cases, a good situation has meant being linked to other places, including resource regions and other cities. Before the advent of the railroad or the automobile, waterways provided some of the most important links among places. Both Eastern and Western Europe illustrate that a location on a major river or one of its tributaries is favorable to city growth.

YOU ARE THE GEOGRAPHER

1. You will need to use the map on page 47 to complete this activity. Begin by using an atlas to correctly identify and label the following Eastern European countries: Poland, the Czech Republic, Slovakia, Hungary, Slovenia, Croatia, Bosnia and Herzegovina, Yugoslavia, Albania, Macedonia, Bulgaria, and Romania. Also label Germany and Austria, although they are not Eastern European. Try to label the country names in such a way that you will have room to write the river names along the rivers and the city names near the dots provided.

2. The five longest rivers in Eastern Europe (excluding Belarus, Ukraine, and Russia) are the Danube, the Elbe, the Vistula, the Tisza, and the Sava. Not all of these rivers are entirely within Eastern Europe. The Danube actually begins in Germany, and the Elbe ends there. Use an atlas to correctly identify and label these five rivers on the map. Notice that the Sava and the Tisza are major tributaries of the Danube. Also identify and label these additional tributaries of the Danube: the Bosna, the Iskur, and the Dambovita, plus two tributaries of the Elbe: the Havel and the Vltava. Highlight all of these waterways with a blue felt pen.

3. Use the atlas to identify and label the Eastern European capital cities: Warsaw, Prague, Bratislava, Budapest, Ljubljana, Zagreb, Sarajevo, Belgrade, Tirane, Skopje, Sofia, and Bucharest. Also label Berlin and Vienna in Western Europe.

4. Which national capitals are located on the Danube River system?

_____

5. Which national capitals are located on the Elbe River system?

_____

6. Which national capital is located on the Vistula?

_____

7. Which capitals are not on any of the five rivers named above? Are they on other rivers (check your atlas)?

_____

Eastern Europe

SCALE

0 — 125 — 250 Miles

0 — 125 — 250 Kilometers

Projection: Azimuthal Equal Area

8. What is the main direction of flow of the Elbe? Into what body of water does it empty? What do you suppose happened to the amount of barge traffic on the Elbe when the "Iron Curtain" came down after World War II, leaving Hamburg on one side and many upstream cities on the other?

_____

_____

_____

9. What is the main direction of flow of the Danube? Into what body of water does it empty? Now look in your atlas for the Rhine River in Western Europe. Also find the Main River, one of its tributaries. Label these on your map and highlight them in green. The Rhine is a major river like the Danube that flows through many countries and cities. Can you think of reasons why the amount of traffic on the Rhine is much greater than that on the Danube, despite the Danube's much greater length?

_____

_____

_____

10. In 1998 the Danube and Rhine River systems were linked when the Rhine-Main-Danube Canal was completed. (The Vistula and Elbe were already linked by canal to the Rhine.) The idea for making this link had been around since the time of Charlemagne (742–814). Look for the German city of Nürnberg in your atlas. It is located on the canal. Draw in the canal on your map. Highlight it in red. What might be the effect of this Rhine-Main-Danube Canal on the cities along the Danube system?

_____

_____

_____

The Soviet "Game of the Name"

Where were you born? St. Petersburg. Where did you grow up? Petrograd. Where do you live? Leningrad. Where would you like to live? St. Petersburg. This joke was published in *The Economist* magazine in 1991. Why is it a joke? Because St. Petersburg-Petrograd-Leningrad-St. Petersburg are all names of a single place, the city founded on the shores of the Baltic Sea in 1703 by Peter I of Russia. St. Petersburg is not the only place in Russia to experience official renaming. Scores of towns and cities (and other features) in Russia and the rest of the Soviet Union had politically motivated name changes during the 75 years of Soviet rule.

In general, *toponyms* (place-names) are mirrors of the societies that choose them. The Soviet use of toponyms demonstrates the wish to influence people's opinions. In various times and places, toponyms reveal their namers' religious beliefs, practical bent, sense of humor, cultural or national icons and heroes, and many other attitudes and traits. Toponyms help turn places into symbols of folk or popular culture.

Soon after obtaining power, the Soviets wiped off the map names that belonged to Russia's czarist past. These old names were replaced with the names of revolutionaries and leaders of the Communist Party and Soviet Union. If those same revolutionaries and leaders later became unpopular, their names were removed and new ones were chosen. For example, in the 1920s and 1930s there were half a dozen cities named for Joseph Stalin. However, in 1961, then-president Nikita Khrushchev changed them all after Stalin's crimes were revealed. Some cities experienced as many as four name changes as a result of the shifting winds of political fortune. In addition to Soviet communist leaders, other categories of people honored with toponyms included revolutionaries and communists from other times and places and those who had made major (and, of course, politically correct) contributions to Russian or Soviet arts and sciences.

YOU ARE THE GEOGRAPHER

The table on page 50 contains 10 cities in Russia and Ukraine. This is just a sample of the many cities that experienced the Soviet "game of the name." Begin by using an atlas to correctly label these on the map on page 51 using their current names. (If you have an old atlas, you may have to look under a previous name.)

Next, use an encyclopedia to look up the underlined names (your teacher may choose to assign one or several of the names to you, rather than have you look up every name yourself). When you are reading about these people, note any deeds that would have made them villains or heroes (or both, at different times!) to the Soviet Union's communist leaders. Pay attention to the pattern of dates of city name changes. They often come in clusters, reflecting some major Soviet ideological shift.

Write about what you learned. Share it in a group discussion. Can you come up with place-names in your state or region that honor exceptional people? Do you know of any that have been changed out of a desire to forget someone's less than honorable deeds?

Selected Russian and Ukrainian City Names

Current city name, date acquired, & person named for	Current city population	Former city name, dates applicable, & person named for	Former city name, dates applicable, & person named for	Former city name, dates applicable, & person named for
St. Petersburg 1991 St. Peter	4,952,000	Leningrad 1924–1991 V. I. Lenin	Petrograd 1914–1924 St. Peter; also Peter I (the Great)	St. Petersburg 1703–1914 St. Peter
Gatchina 1944	81,300	Trotsk 1923–1929 L. D. Trotsky	Krasnogvardeysk 1929–1944	Gatchina 1795–1923
Volgograd 1961 [Volga River]	1,006,000	Stalingrad 1925–1961 I. V. Stalin	Tsaritsyn Before 1925	—
Luhansk, Ukraine 1990	504,000	Voroshilovgrad 1970–1990 & 1935–1958 K. Y. Voroshilov	Lugansk 1797–1935 & 1958–1970	Yekaterinoslav 1795–1797 Catherine II (the Great)
Perm 1957	1,099,000	Molotov 1940–1957 V. M. Molotov	Perm 1780–1940	—
Naberezhnyye Chelny 1988	514,000	Brezhnev 1982–1988 L. I. Brezhnev	Naberezhnyye Chelny 1930–1982	Chelny Before 1930
Engels 1931 F. Engels	183,000	Pokrovsk Before 1931	—	—
Gagarin 1968 Y. A. Gagarin	Not known	Gzhatsk Before 1968	—	—
Nizhniy Novgorod 1990	3,704,000	Gorky 1932–1990 M. Gorky	Nizhniy Novgorod Before 1932	—
Pushkin 1937 A. S. Pushkin	95,300	Petsokye Selo 1918–1937 (means Children's Village)	Tsarskoye Selo 1710–1918 (means Tsar's Village)	—

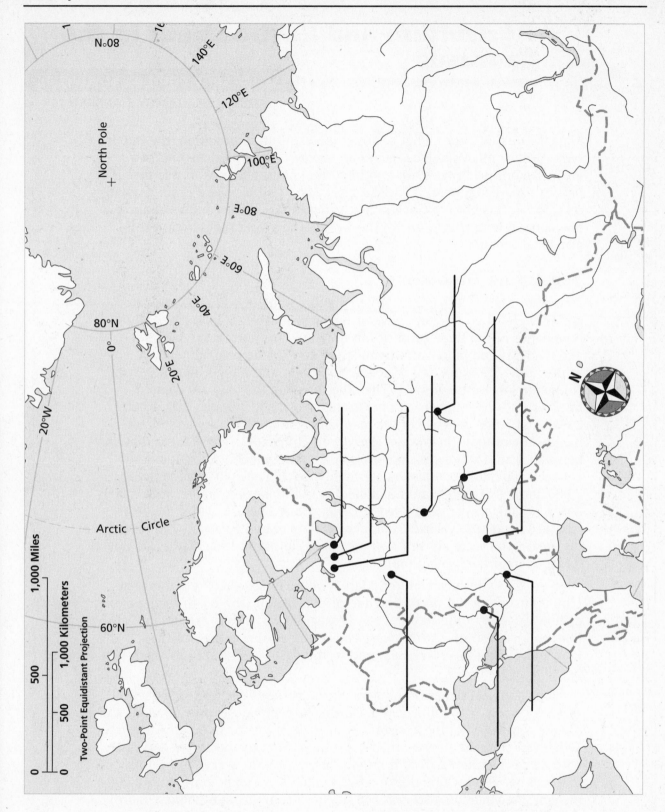

Agriculture and Environment in the Caucasus

GEOGRAPHY FOR LIFE ACTIVITIES

NATIONAL GEOGRAPHY STANDARD 15

Many factors influence what a farmer grows. Cultural traditions are important. Farmers often produce what has "always" been produced in their area. Demand and prices are also important considerations. If a farmer is part of a market system, he or she must ask, "Can I make a profit growing this crop at this place?" Government agricultural policies and assistance play a big role in some parts of the world. The physical environment is also a key factor.

YOU ARE THE GEOGRAPHER

First color Map 1 on page 53. When you are done, it will show six categories of elevation (height above or below sea level). Map 1 shows the boundaries of the three countries in the Caucasus (Armenia, Azerbaijan, and Georgia) and land-water boundaries. Use a blue colored pencil to color in the Black Sea, the Caspian Sea, Lake Sevan, and Mingechaur Reservoir (these are labeled on Map 2). The other lines on the map are contour lines, or lines of equal elevation. On this map, they are at 0, 200, 1,000, 2,000, and 3,000 meters. On the map, areas of land that are less than 0 meters in elevation (below sea level) are numbered 1. Those areas that are between 0 and 200 meters in elevation are numbered 2. Those that are between 200 and 1,000 meters are numbered 3. Those that are between 1,000 and 2,000 meters are numbered 4, and those that are between 2,000 and 3,000 meters are numbered 5. Using colored pencils, color the areas labeled 1, green; 2, yellow; 3, orange; 4, brown; and 5, black. The remaining areas, which are above 3,000 meters in elevation, leave white. Do not forget to color the legend, too.

Now that you have colored Map 1, you can see the Great Caucasus Mountains along the northern edge of this region (the north-facing slope of this range is in Russia) and the Lesser Caucasus and Armenian Plateau in the south. At the west end of the region is the Colchis Lowland, facing the Black Sea, and in the east is the Kura Lowland, facing the Caspian.

Topographical patterns and agricultural patterns are related for several reasons. Topography influences soils, slope, temperature, precipitation, and ease of access. All of these factors in turn affect what can be grown where. When you compare the two maps, remember that temperatures decrease with higher elevations. This is important because there are many crops that will not grow where freezing temperatures are common (above 2,000 meters). Precipitation also varies in this region. In general, there is more moisture in the west and less in the east. Samtredia, on the Colchis Lowland, gets an average of 56 inches of rain per year. Tbilisi, close to the center of the region, gets 20 inches per year. Zyud-Ostov-Kultuk, on the Kura Lowland, gets only 11 inches per year.

Map 1: Contour Map of the Caucasus

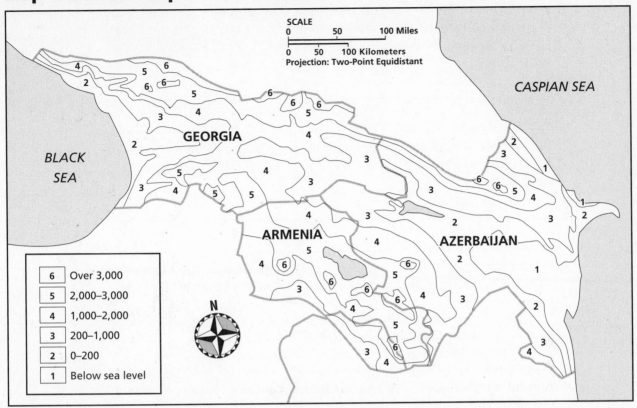

SCALE
0   50   100 Miles
0   50   100 Kilometers
Projection: Two-Point Equidistant

CASPIAN SEA

GEORGIA

BLACK SEA

ARMENIA

AZERBAIJAN

N

6	Over 3,000
5	2,000–3,000
4	1,000–2,000
3	200–1,000
2	0–200
1	Below sea level

Map 2: Agricultural Patterns of the Caucasus

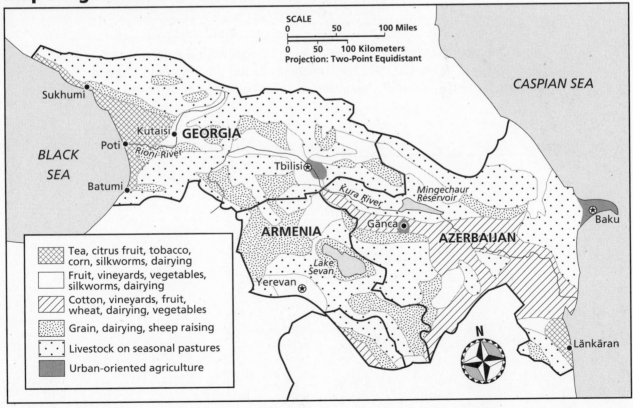

SCALE
0   50   100 Miles
0   50   100 Kilometers
Projection: Two-Point Equidistant

CASPIAN SEA

Sukhumi

BLACK SEA

Kutaisi   GEORGIA
Poti
Rioni River

Batumi

Tbilisi

Kura River

Mingechaur Reservoir

Gäncä

Baku

ARMENIA

AZERBAIJAN

Lake Sevan

Yerevan

Länkäran

N

	Tea, citrus fruit, tobacco, corn, silkworms, dairying
	Fruit, vineyards, vegetables, silkworms, dairying
	Cotton, vineyards, fruit, wheat, dairying, vegetables
	Grain, dairying, sheep raising
	Livestock on seasonal pastures
	Urban-oriented agriculture

Now compare Maps 1 and 2. Then answer the questions that follow.

1. What is raised in the lowland regions of the Caucasus? Why do some of the agricultural products differ between the Colchis and the Kura Lowlands? What do you think Mingechaur Reservoir has to do with agriculture?

_____

_____

2. What types of agriculture dominate in the medium-elevation areas of the Caucasus? How are vines, tree crops, and animals well suited to steep slopes?

_____

_____

3. What type of agricultural use is the most important in the highest elevation parts of the Caucasus? Why are pastures "seasonal" in these areas?

_____

_____

4. Around which places is urban-oriented agriculture found? What is meant by this term?

_____

_____

5. Finish this exercise with a group discussion of agricultural patterns in your locality or state. Can you see the effects of the physical environment on those patterns? Are there examples of urban-oriented agriculture?

_____

_____

Environmental Crises in the Aral Sea Basin

The Aral Sea Basin, just east of the Caspian Sea in Central Asia, is the location of one of the most serious cases of ecological destruction caused by humans in the 1900s. Since the early 1960s, the level of the Aral Sea has dropped drastically. Its area and volume have shrunk. Salinity (saltiness) has increased. Many severe environmental and human problems have accompanied these changes.

DRAINAGE NEAR THE ARAL SEA

The Aral Sea Basin is an example of *interior drainage*. The waters of two major rivers, the Amu Dar'ya and the Syr Dar'ya, empty into the Aral Sea. Water is lost from it by evaporation, not by outflow. Until about 1960, the water balance was in long-term equilibrium. A wide variety of plants and animals lived in the lake and river deltas. Human uses of the Aral Sea and its area included irrigated agriculture, animal raising, hunting and trapping, fishing, and reed harvesting.

In the 1950s, the Soviet Union began to construct large irrigation projects in the basin. For 30 years large diversions, such as the Kara Kum Canal, were built to irrigate crops such as grains, fruits and vegetables, and cotton. Irrigation agriculture had been practiced for centuries in the Aral Sea Basin. The Fergana Valley and areas around the region's ancient cities such as Tashkent and Samarqand are examples. However, the scale of the new diversions was much greater than in the past. Eventually, the Syr Dar'ya and Amu Dar'ya were emptied of most of the water they drained from the Pamir and Tian Shan mountain ranges. The inflow to the Aral Sea was cut to a trickle.

THE RESULTS OF DRAINAGE

The results of this drainage are severe and widespread. The increasing salinity of the shrinking lake killed off all native species of fish. The drop in the fish production destroyed an important commercial fishery that employed tens of thousands of people. Navigation on the Aral Sea was abandoned. Delta ecosystems have been affected by the dryness. Loss of water together with increasing water pollution has wiped out aquatic organisms, including birds. Plant and animal agriculture around the Aral Sea have been damaged by desertification and salinization. Blowing dust and salt from the former sea bottom has reduced crop yields. The local climate has become colder and drier. Serious human health problems, including respiratory and digestive diseases, and possibly cancer, are related to the environmental conditions of the past 50 years.

FIXING THE PROBLEMS

Before the Soviet Union broke up in 1991, it began to address the Aral Sea Basin environmental catastrophe. Now, the newly independent states of Central Asia (Kazakhstan, Uzbekistan, Turkmenistan, Kyrgyzstan, and

Tajikistan) have taken over responsibility for solving the many problems they inherited. Iran, Afghanistan, and China also control parts of the Aral Sea Basin. However, they have not been part of the agreements signed so far by the former Soviet states. These states are receiving help from a wide variety of organizations and governments. Among the most important tasks for the countries involved is to come to an agreement on how to share the water of the Amu Dar'ya and Syr Dar'ya, while leaving enough to repair the environmental damage in the Aral Sea region.

YOU ARE THE GEOGRAPHER

Now use an atlas to locate all of the places underlined in the text. Label them on the map on page 57. Highlight the water bodies in blue, the boundary of the Aral Sea Basin in red, and the country boundaries in yellow.

1. Compute the percentages missing from the table below and fill them in.

2. Which has shrunk more, the lake's area or its volume?

_____

3. About how many times saltier is the Aral Sea predicted to be in 2010 than it was in 1960?

_____

4. Make a graph of the lake's changing level, area, volume, or salinity.

Characteristics of the Changing Aral Sea

Year	Average level, m	% of 1960 amount	Average area, km²	% of 1960 amount	Average volume, km³	% of 1960 amount	Average salinity, grams/ liter	% of 1960 amount
1960	53.4	100	66,900	100	1090	100	10	100
1971	51.1		60,200		925		11	
1976	48.3		55,700		763		14	
1989	39.1		39,718		363		30	
1993	37.1		35,228		294		35	
1998	35.2		30,686		235		45	
2010	31.5		23,731		162		60	

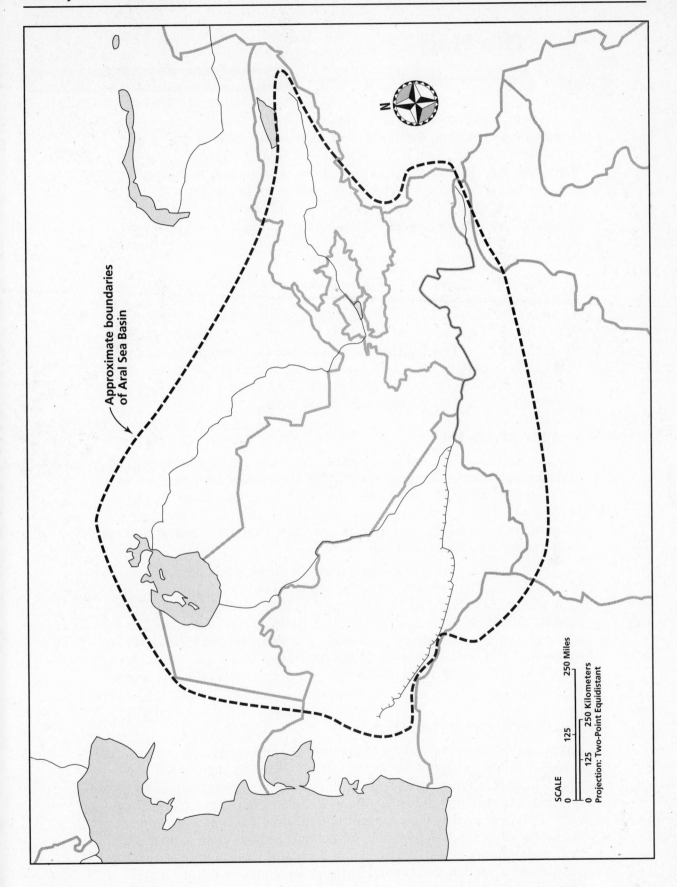

Approximate boundaries of Aral Sea Basin

SCALE

0   125   250 Miles

0   125   250 Kilometers

Projection: Two-Point Equidistant

The Hajj

The Arabian Peninsula, Iran, Iraq, and Afghanistan are part of a large world region where Islam is the most common faith. This region, sometimes called the Islamic World, also includes Albania and Bosnia in Europe, much of northern Africa, the countries of the eastern Mediterranean (except Israel), the former Soviet republics in Central Asia, western China, Pakistan, Bangladesh, Malaysia, and Indonesia. Significant Muslim minorities also exist in India, Singapore, Thailand, the Philippines, South Africa, Yugoslavia, and Great Britain.

The Islamic region is varied in languages, political and economic systems, and elements of everyday life such as dress, diet, and housing. Yet the culture of Islam is an important unifying force. Islam's basic rules are shared by all. These rules are called the Five Pillars of Islam. They are: repeating the basic creed ("There is no God but God, and Muhammad is his prophet"); praying five times a day, facing Mecca; giving charity to the poor; fasting during the month of Ramadan; and making at least one pilgrimage to Mecca. This pilgrimage is called a hajj.

HISTORY OF THE HAJJ

The hajj actually originated before Islam. However, during the prophet Muhammad's lifetime, it became a Muslim observance. Non-Muslims were not allowed to go to Mecca and Medina. Mecca was the birthplace of Muhammad. He spent part of his life teaching in Medina. Before the advent of modern travel, the hajj was a powerful force that brought people from a wide variety of places together.

The hajj always takes place on the eighth through the thirteenth days of Dhu al-Hijjah, the last month of the Muslim year. In recent years the total number of pilgrims has been about 1 million. Participants in the rituals join an enormous crowd. The numbers of accommodations, meals, transport vehicles, emergency personnel, and so on needed for the hajj are staggering. This activity focuses on the geography of the hajj. More specifically, it focuses on the geography of the source countries of hajj pilgrims. How many come? From where do they come, and why?

THE GEOGRAPHY OF THE HAJJ

Unfortunately, recent figures on how many pilgrims travel from foreign countries to Mecca each year are unavailable. However, figures are published on how many people (in general, not pilgrims) from various foreign countries arrive in and depart from Saudi Arabia each year. Column 1 in Table 1 on page 60 lists the countries with at least 1,000 arrivals or departures to or from Saudi Arabia in 1996; column 2 shows how many visitors from each country departed from Saudi Arabia in 1996 (numbers arriving and departing are not the same; departures were chosen for use here). Column 3 lists the total population of each of these countries in 1997.

Column 4 consists of column 2 as a percentage of column 3. Column 5 shows the estimated percentage of each country's population that is Muslim.

These factors—total population size and the percentage that is Muslim—should help you understand something of the pattern of travel between Saudi Arabia and other countries. Distance from Mecca and a country's degree of wealth are also factors in the hajj, but these factors will not be included in this activity. Only about 1 million out of 7 million visitors to Saudi Arabia are hajj pilgrims, but at least some ideas about the geography of hajj travel can be gained from this information.

YOU ARE THE GEOGRAPHER

Answer the following questions using the table, your textbook, an atlas, and other resources (including class discussion) as needed.

1. Which small countries have generated many visits to Saudi Arabia because they are close to it? Why might people from these countries be visiting Saudi Arabia, other than for the hajj?

2. Which large country near Saudi Arabia generated more visits than any other? Why might people from this country visit Saudi Arabia, other than for the hajj?

3. Three North African countries listed in the table each have about 28 million people. The poorest of these three countries generated more than three times as many visitors to Saudi Arabia as either of the others. Which country was that? What might be reasons for this?

4. Why are Iraq's numbers so low? How about the low number for Libya? Hint: What is the political situation in each country?

5. This country has the world's largest Muslim population, but it is fairly poor and far from Saudi Arabia. What country is this?

6. What might be going on with the Philippines? It is only 5 percent Muslim, yet large numbers of visitors went to Saudi Arabia. Why might Filipinos come to Saudi Arabia, other than for the hajj? The same factor explains some of the movement to and from Pakistan, India, Sri Lanka, and Bangladesh, too.

7. Why might Japanese and South Koreans, who are not likely to be Muslim, visit Saudi Arabia? What natural resource that the Saudis have in abundance do Japan and South Korea lack?

8. Which country has a higher percentage of Muslims, India or Pakistan?

9. How does South Africa fit into the picture? Every other African country listed (some of the African countries are categorized as Arab countries) is in northern Africa, where Islam spread by contact diffusion. Hint: Remember the British Empire and how people from different parts of the world moved around within it.

10. How might you go about finding an estimate of how many Muslims from the United States made the pilgrimage to Mecca in a recent year?

Table 1 Foreigners Leaving Saudi Arabia (1996)

Nationality	Number Leaving Saudi Arabia	Total Population (millions)	Percentage of Total Population	Muslim Population
Arab Countries				
Algeria	47,507	29.8	< 1.0%	99%
Bahrain	840,613	0.6	140.1%	100%
Djibouti	1,494	0.6	< 1.0%	94%
Egypt	1,299,576	64.8	2.0%	94%
Iraq	4,187	21.2	< 1.0%	97%
Jordan	381,237	4.4	8.7%	96%
Kuwait	557,176	1.8	31.0%	85%
Lebanon	63,286	3.9	1.6%	70%
Libya	1,001	5.6	< 1.0%	97%
Mauritania	2,451	2.4	< 1.0%	100%
Morocco	49,203	28.2	< 1.0%	99%
Oman	62,178	2.3	2.7%	"Mostly"
Palestine	81,964	5.8	1.4%	14%
Qatar	243,846	0.6	40.6%	95%
Somalia	15,857	10.2	< 1.0%	"Mostly"
Sudan	187,381	27.9	< 1.0%	70%
Syria	358,225	15.0	2.4%	90%
Tunisia	31,833	9.3	< 1.0%	98%
U.A.E.	88,702	2.3	3.9%	96%
Yemen	230,218	15.2	1.5%	"Mostly"
Asian Countries				
Afghanistan	6,624	22.1	< 1.0%	99%
Bangladesh	130,964	122.2	< 1.0%	88%
China	3,006	1,236.7	< 1.0%	3%
India	600,901	967.7	< 1.0%	14%
Indonesia	144,223	204.3	< 1.0%	87%
Iran	101,412	67.5	< 1.0%	99%
Japan	5,899	126.1	< 1.0%	0%
Malaysia	47,552	21.0	< 1.0%	60%
Pakistan	474,677	137.8	< 1.0%	97%
Philippines	270,485	73.4	< 1.0%	5%
Singapore	8,708	3.5	< 1.0%	15%
South Korea	7,172	45.9	< 1.0%	0%
Sri Lanka	89,503	18.7	< 1.0%	8%
Thailand	11,974	60.1	< 1.0%	4%
Turkey	88,564	63.7	< 1.0%	100%
African Countries				
Chad	11,387	7.0	< 1.0%	50%
Ethiopia	31,849	58.7	< 1.0%	48%
Mali	4,062	9.9	< 1.0%	90%
Nigeria	40,793	107.1	< 1.0%	50%
South Africa	9,034	42.5	< 1.0%	2%

What Future for Jerusalem?

Capital cities are important places. They are the seats of government and centers of political and military power. They symbolize national identity. Often, they are important economic and cultural centers, too. People invest huge resources—financial and psychological—in capital cities. What happens when more than one country lays claim to the same city as its capital? This is the situation in Jerusalem.

In this activity, you will read information about Jerusalem and the conflict between Israel and the Palestinians with whom they share land. Then you will explore four possible futures for Jerusalem. Before you read further, look at a map of Israel in your textbook or an atlas. Find the areas of the West Bank (notice that it is on the west bank of the Jordan River), Gaza, and the Negev. Also note the cities of Jerusalem, Tel Aviv, Hebron, Beersheva, Nablus, and Ramallah. Consider also Israel's Arab neighbors, particularly Jordan. The region of Palestine covers essentially the same territory as Israel and the West Bank. The Palestinians are Muslim Arabs who have lived in Palestine for hundreds of years.

JERUSALEM'S HISTORY

King David made Jerusalem the capital of all the tribes of Israel in 1047 B.C. Modern-day scholars suggest that David chose Jerusalem (rather than his own city of Hebron in Judah) to be the capital because all the tribes could share it. Thus, it did not belong to any one tribe. David moved the Ark of the Covenant to Jerusalem, thereby signaling its status as the capital of a united Israel. The Ark of the Covenant was the most important religious symbol of the Israelites.

In the centuries that followed, this country known as Palestine was conquered more than 20 times by outsiders. Under some of these foreign administrations, Jerusalem was made the capital of a surrounding province. In contrast, the Muslim Arab conquerors of the 700s saw Jerusalem not as a political capital, but as a holy city. In the walled Old City, they constructed the Dome of the Rock at a place associated with their prophet, Muhammad. This Muslim sacred site is on the Temple Mount (the Haram al-Sharif to Muslims). The Temple Mount is the location of the Jewish temple that was destroyed by the Romans in A.D. 70. It is therefore sacred to Jews also. The Old City also contains sites sacred to Christians. Jerusalem was in Christian hands for nearly 200 years during the time of the Crusades. It switched back to Muslim control again until World War I, when it was taken by British troops.

JERUSALEM IN THE 1900s

After World War I, Britain was given legal authority to govern Palestine by the League of Nations. In 1924 Britain chose Jerusalem as the capital of this territory. Meanwhile, the British said they would support the plans of the Zionists. Zionists were Jews from around the world who hoped to establish a new Jewish nation-state in Palestine, their historic homeland. The Palestinians objected.

In the beginning, the city of Jerusalem was not a focus of either the Zionists or the Palestinians. Both groups were more interested in the economically vital coastal cities. However, as the two groups put forth their competing nationalist claims to the same territory in the 1930s, Jerusalem became a more central concern.

When Israel declared its independence in 1948, Jordan invaded. At the end of the war the next year, Jordan held East Jerusalem and the West Bank. Israel ended up with quite a bit more territory than what had been proposed in the United Nations partition plan. At the end of this war, Jerusalem was split into Jordanian East Jerusalem and Israeli West Jerusalem. The city was divided by walls and guarded by UN peacekeepers.

As long as the city was divided, many inhabitants were denied access to holy sites that were located across the boundary. Most of the major holy sites of all three religions, including the Old City, were on the Jordanian side. At this point in time, the United Nations wanted to make Jerusalem an international city.

Israeli West Jerusalem was surrounded on three sides by the Jordanian-held West Bank. In spite of this situation, Israel began moving central government functions into West Jerusalem in 1948. Israel officially declared Jerusalem its capital in 1950.

In 1967 Israel won the Six-Day War against its Arab neighbors. Among other things, it took the West Bank and East Jerusalem from Jordan. Israel annexed Arab East Jerusalem and reunited the city. Jerusalem, no longer surrounded by Jordanian-held lands, became even more central to Israel. New investment poured in and the population grew. Some Jewish neighborhoods and suburbs were built on the Arab eastern side. However, because Israel ignored the UN plan to make Jerusalem an international city, no country has recognized Jerusalem as the capital of Israel. Most countries continue to have their embassies in Tel Aviv.

REACHING AN AGREEMENT

The Israelis and Palestinians have been working toward an agreement over Jerusalem. The Palestinians would like an independent country of their own. The old UN plan supported this idea. This Palestinian country would include Gaza (which they already govern without much Israeli involvement), the West Bank, and East Jerusalem. The Israelis have strong objections to this. One reason for their objections is that they have allowed Jewish settlement in the West Bank.

Both Jewish Israelis and Muslim Palestinians believe that their claims to Jerusalem are valid. Christians are also part of the picture. They want to be able to visit the many sites holy to their religion. How can these claims be resolved? One geographer, Chad Emmett, has outlined four possible futures for Jerusalem.

PLAN 1: A UNITED CAPITAL

In this plan, the status of Jerusalem would be as Israel sees it: "Jerusalem united in its entirety as the capital of Israel." The city would remain united under Israeli rule. Palestinian residents of the city (who now make up more than one quarter of the total population) could hold some type of dual citizenship—Israeli and West Bank. A new West Bank (Palestinian) capital complex might even be located in East Jerusalem. A city council would make decisions on region-wide issues. Jewish and Palestinian community boards would make decisions on things like local land use, schools, and parks. The holy sites of each religion, clustered in and around the walled Old City, would have a special status, guaranteeing access to believers. This future recognizes the importance of Jerusalem to the Palestinians, but leaves the city under Israeli control.

PLAN 2: A FORMER CAPITAL

The Israelis and Palestinians would both choose capitals other than Jerusalem. Israel could move its capital back to Tel Aviv, where it was in the earliest days of the country's existence. It could select Beersheva, in order to encourage further development of the Negev (southern) region of Israel. Nablus, Ramallah (which is just north of Jerusalem and already has some Palestinian government offices), Hebron (King David's old city), or Gaza (which is the seat of the emerging Palestinian government now) are possible Palestinian choices. If Jerusalem were not anyone's national capital, it might be easier for the groups to compromise.

PLAN 3: AN INTERNATIONAL CAPITAL

This future emphasizes the idea that Jerusalem belongs not just to Israel and Palestine, but also to all for whom the city has spiritual meaning. Jerusalem could be separated politically from Israel and the West Bank/Palestine and governed by an international group or a number of religious bodies. Another option would be to have Israel govern the Jewish parts of the city and Palestine the Arab parts. The holy places would be internationalized and governed by outsiders.

PLAN 4: A SHARED CAPITAL

Some ideas for sharing Jerusalem echo the situation of 1949–1967. During this time Jerusalem was divided by physical barriers into two parts. This plan is not popular. Other ideas include allowing Jerusalem to be the national capital of two countries exercising joint sovereignty (power) over it. There would be an Israeli mayor and city council for Israeli residents and a Palestinian mayor and city council for Palestinian residents. Together the councils could make decisions affecting the entire city. A joint authority could manage the holy sites. Everyone would have access to the entire city. There could be efforts to merge some scattered Jewish and Palestinian neighborhoods if desired.

YOU ARE THE GEOGRAPHER

In small discussion groups, list the pros and cons for each of the four possible plans. Think not only about what is desirable or undesirable about each plan, but also about what is practical or impractical. After discussion, decide which future for Jerusalem you think is the best choice and why. Share your group's work in a class discussion or in a written summary.

The Camel

"The blessings of this world, until judgement day, are tied to the forelocks of our horses; sheep are a benediction; and the Almighty, in making animals, created nothing preferable to the camel."—Muhammad, the Prophet of God

"There is undoubtedly something about the camel which is at first repulsive, not merely in its odour, but in its peculiarly reptilian appearance. He has been aptly described as resembling a cross between a snake and a folding camp-bedstead."—L. M. Hira, 1947

The camel is of immense importance to the nomadic peoples whose lives and cultures depend upon it. Camels appear in many Arabic proverbs and poems, where the animal's value and beauty are highly praised. A 1953 study of a dialect spoken by a camel-herding people in Mauritania identified almost 700 words relating to camels. These words described all aspects of the camel including its physical characteristics (40 words had to do with different coat colors alone), behaviors, temperaments, and ages. If a single creature provided most of the necessities in your life, your vocabulary relating to it would be very rich, too.

In this activity you will write a short story or poem set in northern Africa incorporating some knowledge of the camel and its impact on the human societies and landscapes where it remains important. Background information is provided here, but feel free to supplement it with additional material.

WHERE ARE THE CAMELS?

As of 1998 there were just over 19 million camels in the world, up from 11 million 50 years earlier. Of that 19 million, 14.6 million or 75 percent were located in Africa. Nearly all the camels in Africa are in the northern parts of countries in the Sahara and the Sahel. Kenya is the southernmost African country in which the camel is important. There, higher humidity (which interferes with camel breeding) and the tsetse fly (which carries diseases that kill domestic livestock) become problems. The remaining 25 percent of camels are mostly in Asia, particularly in the Middle East, Pakistan, and India.

THE ROLE OF THE CAMEL

R. T. Wilson, in his 1984 book, *The Camel*, divided countries with camels into four categories. These categories were based on the importance of the camel in the livestock economy of each country. Essentially, he figured out how important camels (or more exactly, total camel biomass) were compared to cattle, sheep, goats, horses, mules, donkeys, buffaloes, and camels all taken together (*total* "domestic herbivore biomass" or DHB). In Africa, the countries of Nigeria, Senegal, and Burkina Faso fell into his Category 1. Category 1 meant that in this country camels were present in considerable numbers, but they made up less than 1 percent of DHB. In those countries, there are no large ethnic groups for whom the camel is important culturally.

In Category 2 countries, camels contributed between 1 and 8 percent of DHB. In Africa, Egypt, Libya, Algeria, Morocco, Mali, Ethiopia, and Kenya were Category 2 countries. The camel is not central to the way of life of many people in any of these countries, but it is central to at least one group. Examples include the Tuareg in Mali and Algeria, the Bedouins in Egypt, the Afar and Oromo in Ethiopia, and the Somali in Kenya.

Tunisia, Niger, Chad, and Sudan are the African countries in Category 3. Category 3 countries include those in which camels make up 8 to 20 percent of DHB. Here, considerable segments of the population have cultural affinities with the camel. It has become important as a provider of meat, transportion, and some milk.

The final category, Category 4, includes countries in which the camel makes up more than 20 percent of DHB. It is of overwhelming importance to the culture, economy, and ecology of these places. Mauritania, Djibouti, Somalia, and Western Sahara are in this category. Camels made up a whopping 89 percent of total DHB in Western Sahara according to Wilson and 54 percent in Somalia. In 1998 Somalia had more than 6 million camels, almost one third of the world's total!

HISTORY OF THE CAMEL

Evidence suggests that the camel was domesticated in southern Arabia starting in the period 3000–2500 B.C. The reason for the domestication was milk. Diffusion across the Red Sea to Somalia occurred early, where the camel's primary use is still as a milk animal. Meanwhile, back in Arabia, the camel's potential as a pack animal was recognized. It came to be used in the incense trade from southern Arabia to northern Arabia and Syria. The next major step in the rise of the camel to importance was the development of the North Arabian camel saddle. This saddle allowed a rider to sit up high over the camel's hump. The much greater stability of this saddle meant the rider could use a spear and sword as weapons while riding. Before this saddle the desert tribes used bows and arrows in combat. Their greater military success brought them greater political and economic power. They and their camels gradually became accepted by settled Middle Eastern society.

The camel diffused across the southern Sahara from Arabia. In North Africa various shoulder saddles (which sit in front of the hump) were developed. These saddles proved the best of all for riding. The Berbers, a non-Arabic nomadic people, adopted the camel and used it as a milk animal and as a riding animal. However, the major caravan trade already characteristic of the Middle East did not develop in North Africa until the Muslim conquests of the A.D. 700s.

One of the most remarkable consequences of the success of the camel was the disappearance of all types of wheeled vehicles from most Arab lands. This disappearance occurred between A.D. 300 and 500. Wheeled vehicles did not make a major return until the Europeans arrived in the 1800s. Governments invested in bridges and caravanserais (inns designed for overnight stops by camel caravans). They did not invest in roads because they were expensive and unnecessary for camels.

In addition to a lack of roads, the camel's other major influence on Middle East and North African landscapes can be seen in the region's cities. Pre-1800s cities have narrow, winding streets, blind corners, and a generally mazelike character. Why should the streets be straight and wide, when no wheeled vehicles, only people and camels, ever walked them? As late as 1845, the width of a new street in Cairo was determined by the combined width of two loaded camels.

THE CHARACTERISTICS OF THE CAMEL

What are the physical traits of the camel that make it so well suited to the dry lands of North Africa and the Middle East? Its broad, flat, thick-soled cloven hoofs do not sink in the desert sand. It can go for several days without water and then can drink large amounts quickly. It has various water-conserving mechanisms. Its nose is designed to keep out flying sand. It can eat a variety of desert plants. Toughened skin pads on various parts of its body protect it from the hard desert floor when it lies down.

With these characteristics, is there any wonder the camel became the herd animal of choice for nomads in the most extreme deserts of the world? Most camels were kept for milk. Breeding camels for other uses, particularly for transport, also supported nomadic groups. Because of limited vegetation in any one place in the desert, camel herders had to be nomadic, following the appearance of new vegetation.

THE ROLE OF THE CAMEL TODAY

The decline in demand for camels by nonnomads, together with modern governments' efforts to settle (and therefore control) nomadic people, has meant that the camel is no longer present in some countries. Morocco and Libya have many fewer camels today than they had 50 years ago. Egypt experienced a modest decline during the same period, while the numbers in Algeria, Ethiopia, and Tunisia have held fairly steady. In all the other camel-producing countries of North Africa, their numbers have actually increased in the past 50 years.

These different outcomes for the camel are the result of many factors. The historic role and significance of the camel, changes in economic development, and differences in government policies have all affected how the camel has done in various countries. The ability to find new uses for the camel (for example, camels as meat providers for nonnomads) have also affected camel numbers.

YOU ARE THE GEOGRAPHER

Write a short story or poem about a camel or camels in North Africa. Incorporate some of the factual information included here and in any other sources you may find. If possible, illustrate your work with a drawing of a camel. Much of the ancient historical geography of the camel has been traced through artifacts with pictures of camels on them. But apparently the camel is hard to draw: archaeologists sometimes cannot decide if a picture is of a camel or some other animal!

Patterns of West African Migration

Africa has a long history of human habitation and migration. African migration involved one group of people displacing another or moving into uninhabited territory. For example, 5,000 years ago waves of Bantu people from present-day Cameroon and Nigeria spread eastward and southward. They eventually reached present-day South Africa. Other groups were pushed out of the way as they expanded.

EUROPEAN INFLUENCES

Patterns of African migration changed when Europeans began arriving. The centers of activity in pre-European West Africa were cities located in the savanna belt. These cities controlled the trade routes between the coastal forests and the interior dry lands. When the Europeans came, the focus of economic activity shifted to the coast. The first West African migration begun by the Europeans was the forced migration of millions of slaves to the Americas. West African coastal states flourished by selling slaves captured in the interior to European traders.

Later, Europeans began using other resources in West Africa. They set up colonies whose function was to provide raw materials from farms, forests, and mines for industries back home. A great deal of labor was needed to produce these materials. French and British colonial administrations used forced labor in the region to work on plantations and to build roads, railroads, and seaports. The labor migrations set in motion by this colonial economy continue even today. People move within their own countries, as well as internationally, to job-rich regions.

WHERE THE JOBS ARE

As early as the late 1800s, people from interior West Africa migrated seasonally to Senegal to work in its peanut-growing areas. In subsequent decades, cocoa plantations in southern Ghana and Côte d'Ivoire attracted immigrants. Additional crops, including coffee, cotton, palm oil, bananas, and pineapples, drew workers to the tropical coastal zone. Areas of timber and rubber production were also a magnet. Mines producing gold, diamonds, bauxite, manganese ore, iron ore, and phosphates provided jobs for other migrants. As the 1900s wore on, cities with their many service jobs became other important destinations.

CHANGING NEEDS

It is not only a "pull" of jobs in a more prosperous economies that triggers migration. The "push" of little economic opportunity at home is also important. The interior countries of West Africa—Mali, Burkina Faso, Niger, and Chad—include large sections in the Sahel and the Sahara. Traditional pastoral ways of life, particularly in times of drought, cannot support the numbers of people now in those areas. Hundreds of thousands have migrated to cities in their own countries or to coastal neighbors.

An important feature of West African migration has been its circular nature. Circular migration means that many migrants only spend part of the year (or perhaps a year or two) away from home in another region or country and then they return. While this practice has become less true over time, many people still retain ties to their home village. They will still visit it regularly, and may plan to retire to it. Another important connection between many migrants and their home villages is the remittances (money) they send back to family members still there.

YOU ARE THE GEOGRAPHER

The table below displays data on foreign nationals (citizens of other countries) who were living in nine West African destination countries in 1975. You are going to translate the data from this table onto the map of West Africa on page 69. Arrows will connect countries of origin and countries of destination. Each arrow's width will be in proportion to the number of foreign nationals in one country that came from another country.

First label all the countries on the map. Next, create a legend to explain your arrows. Ignore any number in the table less than 10,000 (it represents too small a percentage to map). Find the largest number that you will be mapping. This is the number of foreign nationals from Burkina Faso who were living in Côte d'Ivoire. Decide on an appropriate width of arrow for this number, taking into account the sizes of the countries on your map. One inch would be about right. Then for all the other flows (numbers of foreign nationals), figure out how wide their arrows should be. For example, if you assign 726,200 the width of one inch, then the appropriate width for 180,200 (the number of Guineans in Senegal) would be 180,200 divided by 726,200 or 1/4 of an inch.

Once you have calculated the arrow widths (some will be very narrow, like an ordinary pen line), draw the arrows on the map. Use a bright color. For your map key you will need to show a variety of arrow widths and the number of migrants each represents. Put a title on your map.

Foreign Nationals by Country of Nationality (Origin) and Country of Destination, 1975

Country of Nationality	Country of Destination									
	Ghana	Côte d'Ivoire	Burkina Faso	Senegal	Sierra Leone	Togo	Liberia	Gambia	Mali	Total
Ghana	NA	42,500	17,300	1,000	4,600	30,000	6,600	0	0	102,000
Côte d'Ivoire	18,300	NA	44,400	1,400	0	0	1,500	0	7,900	73,500
Burkina Faso	159,300	726,200	NA	13,700	800	8,000	0	0	47,700	955,000
Senegal	100	19,200	2,100	NA	0	0	200	25,300	11,500	58,400
Sierra Leone	3,000	400	800	8,000	NA	0	4,800	400	1,000	11,500
Togo	244,700	12,100	2,900	0	0	NA	100	0	0	259,800
Liberia	4,600	3,400	700	700	11,000	0	NA	300	1,000	21,700
Gambia	100	100	100	45,600	3,400	0	0	NA	1,000	50,300
Mali	13,400	348,500	21,800	28,900	0	0	1,400	5,500	NA	419,500
Guinea	0	105,800	0	180,200	41,000	0	25,400	17,000	24,100	396,100
Nigeria	55,500	49,600	2,000	0	7,300	0	1,700	0	0	116,100

West African Migration

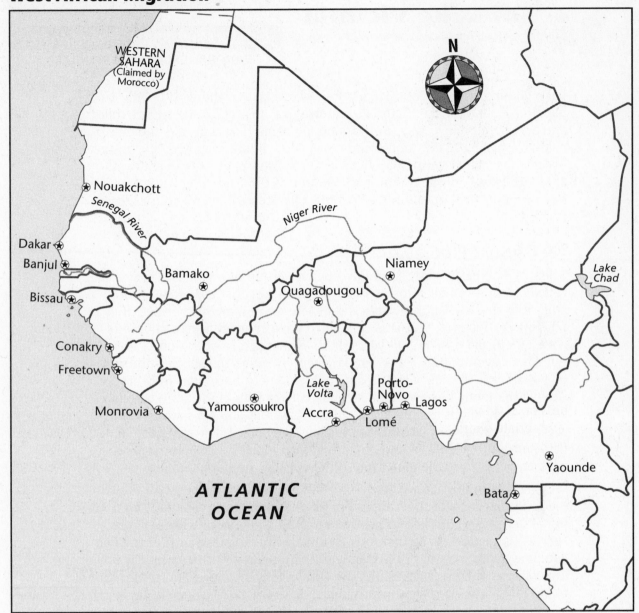

The Disturbed Ecosystem of Lake Victoria

GEOGRAPHY FOR LIFE ACTIVITIES

NATIONAL GEOGRAPHY STANDARD 8

Lake Victoria is the second-largest freshwater lake in the world. Only Lake Superior is larger. Lake Victoria's waters cover nearly 27,000 square miles and are shared by Uganda, Kenya, and Tanzania. For centuries the lake's fish have been a major food source for people living around it. Europeans became fascinated with Lake Victoria when John Speke discovered it in 1859. He later confirmed that it is a source of the Nile River, thus answering a geographical question that people had been asking for thousands of years.

AN UNUSUAL FISH

Scientists are particularly fascinated by one group of organisms in the lake called cichlids—small colorful spiny-finned fishes that are often used as aquarium species. Until recently there were more than 500 species of cichlids in Lake Victoria. These all appear to have evolved from a couple of ancestors in the past 14,000 years or so—an amazing explosion of life in a twinkle of evolutionary time. The key way in which the species differ from one another is in their jaw anatomy, which determines what they eat. Among the foods these various cichlids eat are fish, crabs, prawns, insects, zooplankton, snails, plants, algae, phytoplankton, and parasites. Within these food categories, cichlids are further specialized. For example, some fish-eating species are "snout-engulfing paedophages," which means they suck embryos from the mouths of other species' brooding females.

Changes in the natural ecosystem of Lake Victoria began with the coming of Europeans and the human population growth it triggered. The traditional fisheries of the lake depended most on a couple of native tilapia species (large, mostly herbivorous cichlids). Growing demand for the fish combined with the use of more efficient European-style nets brought about the collapse of these species. So, the British introduced several exotic (non-native) tilapia species from nearby Lake Albert. The introduced tilapia eat a wider range of foods than the native tilapia and are better able to survive in an ecosystem under stress. Still, the cichlids continued to thrive.

Then in 1954 the Nile perch was introduced to Lake Victoria in the hope of converting these little bony fishes into something suitable for the dinner table. The fish-eating Nile perch can reach six feet in length and more than 400 pounds in weight. For a quarter-century, its impact was limited. In 1979 the cichlids still made up 80 percent of the lake's fish biomass, its traditional proportion; the Nile perch made up less than 2 percent.

A SUDDEN CRASH

Then came the shock. Within two years, the Nile perch made up 80 percent of the lake's fish biomass, and the cichlids dropped to 1 percent. Scientists discovered that 200 species had disappeared entirely—the largest single episode of extinction ever recorded. A native zooplankton-eating sardine species was doing fine, perhaps helped by a different reproductive strategy than the cichlids. However, scientists predicted that the perch population would not be able to survive on sardines alone and that its population would also crash.

Meanwhile, the region's governments concentrated on helping fishermen exploit the perch. Because of its size, new equipment and processing techniques were needed. The giant, oily monsters could not just be laid in the sun to dry, unlike the tiny fish of the past. So, with good intentions, outsiders, including foreign investors, were brought in to supply bigger fishing boats and to build modern processing plants. They were hugely successful. The processed fish (many frozen) were sent off to urban markets in the region and to international markets beyond. Few local people can afford the fish anymore. The native peoples are able to use scraps—such as heads and tails—from the factories. There are concerns that people in the region are no longer getting enough protein since so much is being exported.

OTHER IMPACTS

Meanwhile, the lake had a bigger problem than the overabundant Nile perch. Large-scale deforestation of the lake basin had begun years earlier. European settlers had cleared the land for plantations. With more farming came more soil erosion and more fertilizer. This resulted in more plant nutrients in the lake. Sewage, runoff from livestock operations, and industrial wastes such as sugar mill residues also contributed to increased levels of nitrogen and phosphorus. Mats of blue-green algae, which consume these nutrients, spread. The population of algae-eating cichlids that might have kept it under control had already collapsed from overpredation by the Nile perch.

The water became murkier, and some surviving species of cichlids crashed, because they needed to be able to see in order to find mates. Now they could not reproduce. Every so often, rotting algae used up most of the oxygen in the lake. At those times, huge numbers of all species of fish suffocated, fell to the bottom of the lake, and decayed, further robbing the lake of oxygen. By the early 1990s, 50 to 70 percent of the lake's water was anoxic (severely lacking oxygen) year-round.

Still, some organisms did well. Prawns, snails, and midges, which need little oxygen, thrived. They fed on the decayed matter at the lake bottom. The cichlids that specialized in eating these foods were gone. The Nile perch now turned to eating prawns directly. This is how it has avoided the crash predicted by scientists.

A NEW ENEMY

About 1990 another invader, this one a plant, appeared in Lake Victoria. The water hyacinth floated down the Kagera River from Rwanda and into the lake. It is native to the Amazon Basin and was introduced to Africa as an ornamental flower in the 1800s. The plant grows very fast and is probably the worst tropical aquatic weed in the world. At home, pests and diseases keep it in check but abroad, given enough nutrients, it takes over. The shallow waters of Lake Victoria were ideal for the water hyacinth. By 1996, 90 percent of the lake's shorelines were choked by the stuff. All the lake's fish breed along these shorelines. Breeding becomes more difficult as the water hyacinth blocks light and removes oxygen from the water.

The water hyacinth grows so thick that fishing boats cannot get through it to reach open water. Nets get tangled in it, and motors get clogged. It obstructs the big pumps that supply Uganda's capital city, Kampala, with water. It plugs the filters of the region's main hydroelectric generating station and of manufacturing plants. It harbors snakes, snails, and mosquitoes. The snails are hosts for the parasites that give people a serious disease that causes blood loss and tissue damage. The mosquitoes transmit malaria. Both of these diseases are being seen more frequently in the Victoria Lake basin.

SOLVING THE PROBLEM

What are the solutions to Lake Victoria's ecological problems? The possible responses to the hyacinth are costly, take a long time, and/or involve poisons. Poisons might get into the perch and reduce its marketability. Now that the food chains of the lake are so disrupted it would take enormous resources to restore the lake's ecosystem. Fortunately, a wide variety of cichlids and the native tilapia are being preserved in aquariums around the world. If conditions should ever permit their reintroduction, they will be available. However, a return to the lake's natural ecosystem will not happen soon. As long as the perch fishery is doing so well, the governments will not interfere with it.

The perch fishery is a key economic resource for the 30 million people of the Victoria Lake basin. There will be widespread suffering if it collapses. Most experts foresee at least a decline. Ecosystems like Lake Victoria's that have been oversimplified generally do not last long. Any decline in a rural industry like fishing tends to increase migration to nearby cities. At this point, it is impossible to know if such a decline would provide an opportunity to return the lake's ecology to something more like that of its past. In any case, the story of Lake Victoria can serve as a warning to the managers of Africa's other great lakes.

YOU ARE THE GEOGRAPHER

Now, design a visual presentation of the story of Lake Victoria's decline. This should be a series of drawings like a cartoon strip. Be sure to picture the changing food chains, the sequence of events leading to the anoxia, the hyacinth invasion, and the problems faced by humans because of ecological deterioration.

Conservation of Forest Resources

Tropical rain forests cover about 7 percent of Africa's land area. African rain forests represent about one fifth of the world total, with another fifth in Asia and three fifths in Latin America. Estimates suggest that Asia has lost about half of its rain forests, Latin America about one fifth, and Africa two thirds. The two main areas of rain forest in Africa are West Africa and Central Africa. Depletion of the tropical rain forest in West Africa is occurring at an annual rate of about 2.1 percent and in Central Africa at 0.6 percent. The rate of loss in West Africa is higher than that of any other region in the world.

Due to agricultural and urban expansion, there are only a few large tracts of tropical rain forest left in West Africa. The situation is different in Central Africa (defined here as Cameroon, Equatorial Guinea, Gabon, Congo, Central African Republic, and the Democratic Republic of the Congo). Although reduced along its edges and the banks of its rivers, the rain forest in Central Africa remains a vast, fairly continuous expanse. Nearly four fifths of Africa's rain forest is here, half in the DRC alone.

Africa's tropical rain forests contain diverse ecosystems. More than half of all Africa's plant and animal species are found in them. The variety of fauna is by far the highest in Africa. It includes 53 primate species, 84 percent of the continent's total. They range from the 1.5-pound dwarf galago (bush baby) to the 350-pound gorilla. The variety of flora, estimated at more than 8,000 species, is rivaled in Africa only by that of the Mediterranean-climate Cape region in South Africa. Many species in Africa's tropical rain forests are endemic, which means that their distribution is restricted to a single region and they are found nowhere else.

WHAT IS KILLING THE RAIN FOREST?

Depletion of the rain forest comes from two processes, (1) clearance for agricultural expansion and fuelwood gathering (for household use) and (2) commercial logging. Overall, losses to agricultural expansion/fuelwood gathering are much greater than those to commercial logging. However, the relative importance of the two varies from country to country. For example, logging is more important in Cameroon and Gabon than in the DRC, where the vast interior is less accessible. It is important to realize, too, that forest losses to logging and farming are not unrelated. The roads that are built by loggers also serve to open an area up to agricultural expansion.

POPULATION GROWTH

What is behind agricultural expansion into forested land? The first explanation is population growth. Central Africa (the countries named above plus Angola and Chad), with 94 million people, is growing faster than any region in the world. Growing at 3 percent per year, the population will double in just 23 years. The average woman has 6.3 children, again more than in any other world region. This situation is tied to a number of factors.

Traditionally, large families bring social prestige and spiritual security (because ancestors are revered). Economics are important. Women do much of the farming as well as water fetching and wood gathering, and one of the few ways they can reduce their own labor is to provide children to help them. Also, land may be allocated to families according to their size. More children will therefore mean more land. A general lack of economic development and educational opportunities (particularly for women) means there are few ways of earning a living other than farming.

Population growth often results in the intensification of agriculture (producing more on the same amount of land), rather than expansion into new areas. However, if land is free or very cheap, it makes sense for a farmer to use more of that resource and less of other inputs (such as labor or money, which would be required to intensify production). Moreover, farming new lands on a regular basis is a traditional part of subsistence agriculture in Africa (and elsewhere in the tropics), because of the rapid decline of soil fertility that comes with cultivation. Shifting cultivation is a sustainable system under conditions of low population pressure. Now, however, population pressure is leading to falling yields and soil degradation, since farmers are forced to shorten the time that land is allowed to lie fallow and recuperate. In countries where there is still abundant land (such as the DRC), people are moving into forested areas not previously cultivated.

LAND TENURE

Land tenure refers to a society's system of owning and using land. Many Central African traditional societies have community land ownership. Individuals and families are assigned plots on a semipermanent basis. Governments and aid donors have interfered with this arrangement, thinking that if farmers did not own their plots outright, they would not be motivated to invest in the land. When governments have taken over ownership of the land themselves, one of two things has tended to happen. In cases where the government kept title to the land, no one felt responsible for it, so people abused it and made even fewer efforts to improve it than under the old communal system. In cases where the government distributed the land to individual private owners, most of it went to a wealthy few, so that small farmers received very little. The wealthy few were not interested in careful management of the land to improve its output over the long run. Neither strategy has resulted in investments that lead to higher productivity and thus to a reduced need to expand into the forests.

FUEL

The main household fuel in Central Africa is wood. Demand for wood has grown along with the population. Traditionally, wood was so abundant that it was treated as a free resource and was taken from land to which everyone in a village or town had access. This system worked well when populations were small. Now it is contributing to forest destruction. Another forest resource that many Central Africans rely on is bush-meat (meat from wild animals). Population growth is putting growing pressure on this resource too as people living in the forest or at its edge increasingly hunt to meet not only their own needs but also growing urban demands.

COMMERCIAL FORESTRY

Commercial forestry practices vary from place to place within Central Africa. In more accessible locations it may be economical to cut a wide variety of species and sizes of trees. In remoter areas, only a few prime trees may be taken because high transportation costs mean that less valuable trees are not worth cutting. Unfortunately, nearly every Central African country reports that commercial logging is not given adequate oversight by government forestry departments. Salaries of forestry officials are so low that they are easily bribed to overlook abuses. Governments are so desperate for foreign investment that they tend to undervalue their forests. By selling the rights to cut for too little, they are not collecting enough money to manage the forests properly and so are contributing to their demise. By not requiring firms to reinvest profits to achieve long-term sustainability, they are dooming the forest.

There has been some government movement toward requiring foreign logging firms to do a significant share of lumber processing before export occurs. This means the African countries get more value for their wood and more jobs in the processing sector for their people. Still, it is mostly unprocessed logs that leave Central African wharves.

CONSERVATION

In the face of all the pressures on the rain forest, Central African countries have taken action to conserve pieces of it in protected areas. About 7 percent of the remaining rain forest is now protected, and the DRC contains the world's largest rain forest park—the 22,000-square mile Salonga National Park. Studies have shown, though, that many unique forest communities are not yet protected. Additionally, many designated areas are not adequately managed. Poaching, mining, road construction, logging, and agriculture occur in these areas. Park management authorities do not have the money or people to prevent many of these activities. Also, some protected areas suffer from lack of local support because local people have to bear the costs of crop and livestock loss to animals within the reserves and are not able to use parklands as they did in the past. On the other hand, there are examples of protected areas that allow local people to follow traditional forest pursuits such as sustainable hunting and gathering that are compatible with nature conservation.

YOU ARE THE GEOGRAPHER

There are many problems facing the forests of Central Africa, including population growth, agricultural expansion, land tenure reform gone awry, fuelwood cutting, hunting, weak supervision of commercial forestry, and an inadequate system of protected areas. All of these problems are made worse by the region's recent history of political instability, civil war, and government corruption. Suppose you became prime minister of one of the countries of Central Africa. Explain in a short essay how you would address these problems. What problem would you attack first? Who would you involve in finding and implementing a solution? What background information would you need? Where could you find the necessary financial resources? Would international, national, or local programs be most effective?

Apartheid

Apartheid (an Afrikaans word that means "apartness") was the system of racial segregation that existed in the Republic of South Africa from 1948 until 1992. Racial segregation and white supremacy were well established before then, but in 1948 the government began legally creating apartheid. Every person in South Africa was classified as black (also referred to as African or Bantu), white, Coloured (people of mixed black, white, and Malayan descent), or Asian (mostly of Indian ancestry, but some Chinese). The efforts to keep these groups apart operated on three scales. "Grand apartheid" was a plan to create "homelands" for blacks on a national scale. This plan would then leave the rest of the country for the whites. "Urban apartheid" focused on creating separate living and business areas for each racial group within cities. "Petty apartheid" involved avoiding any contact between whites and people of other races in the course of daily life.

GRAND APARTHEID

In the 1940s the South African population was only 20 percent white (today it is 10 percent). To make sure that minority whites controlled South Africa, separate black homelands were created. Blacks would be citizens of these homelands and would have no political rights in white South Africa. This policy essentially made all blacks foreigners in their own country.

Rather than create one large (and therefore more powerful) black nation-state, the South African government pursued a "divide and conquer" policy. Blacks were divided into 10 nations based on language differences. Each nation was given a territorial state (homeland). All members of that nation, even if their families had lived in some other part of South Africa for generations, were assigned to the national homeland.

These black states or homelands made up only about 14 percent of the total land in South Africa, while blacks represented more than 66 percent of the total population (now 75 percent). The cities, mining and industrial areas, and best farmlands were reserved for whites. Between 1960 and 1980, 3.5 million black Africans were removed involuntarily from their homes and relocated to their designated homelands. The large numbers of people pushed into small areas caused major environmental problems. Social conflict within the homelands was common. Poverty was widespread.

URBAN APARTHEID

The South African government controlled the movement of black people outside their homelands. Blacks were required to have a work permit to work outside their homeland. The number of permits issued was determined only by the needs of the white economy. Often, the permits did not allow workers to bring their families. Once in the cities, most black workers had to live in crowded all-black areas located on the fringes of the city. They were subject to strict passbook regulations and curfews.

The Group Areas Act of 1950 was aimed at creating cities that were totally racially segregated. Under this law, every city and town was divided into districts that could be occupied by only one racial group. Business districts were treated in the same way. People living in a zone assigned to another race were forced to leave. Some exceptions were allowed. For example, black or Coloured servants were allowed to live in white neighborhoods, but only in limited proportions.

PETTY APARTHEID

Petty or personal apartheid tried to eliminate any contact between whites and nonwhites. Marriage between a white person and a nonwhite person was against the law. Parks, beaches, places of entertainment, restaurants, and hotels were assigned to a single racial group. Whites and nonwhites could not share a meal or drink together in public unless they had a special permit. Schools and universities, churches and cemeteries, and trains and buses all were segregated. Where whites and nonwhites shared a single building, such as a post office, separate entrances and counters were provided. Facilities for nonwhites were never equal to those for whites, nor did the law say that they should be.

THE END OF APARTHEID

Individuals and groups both inside and outside South Africa condemned apartheid. Many protests were crushed and protesters jailed. Some apartheid laws were relaxed in the late 1970s and 1980s. Many people in the international community spoke against apartheid. Many countries refused to trade with South Africa. These trade embargos caused serious economic damage to the country. In the United States, businesses with investments in South Africa were pressured into leaving.

In 1991 the apartheid laws were repealed. The first South African election in which all races participated was held in 1994. The African National Congress, a black party, came to power.

YOU ARE THE GEOGRAPHER

You will now write an essay analyzing the effects of apartheid. Study the maps on page 78. Both maps are of the city of Port Elizabeth in Eastern Cape Province (note that the city center is near the piers on the maps). The maps display the population distribution by race in 1960 and 1990. Write a paragraph describing what you see on the map of 1960. Then write a second paragraph describing the changes that occurred by 1990.

Next, consider racial segregation in South African cities compared with that in U.S. cities. How was apartheid similar to and different from racial segregation in U.S. cities? Remember that every southern state passed Jim Crow laws in the decades after the Civil War. Even northern states had some discriminatory laws. Think about the ways that blacks in the United States were restricted by custom and by law.

Also consider in general why a group in power would find spatial separation from other groups so important in maintaining its power. Summarize your thoughts on these questions in the third paragraph of your essay.

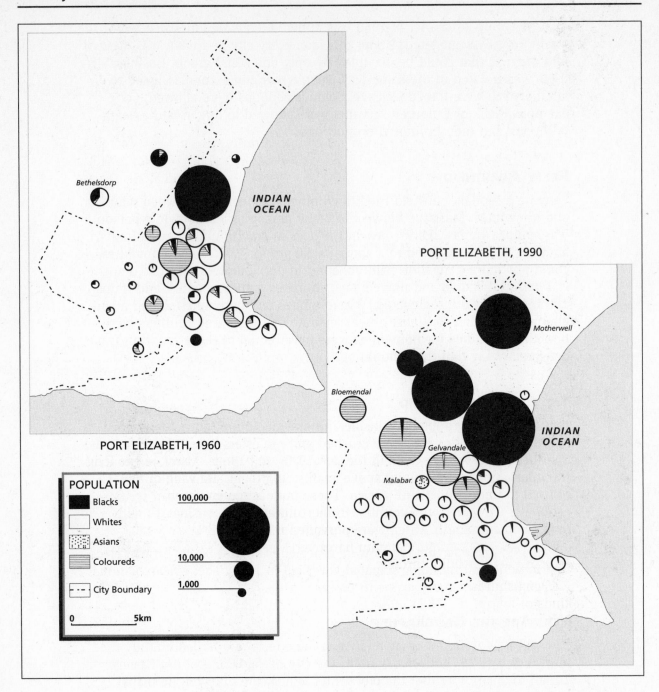

PORT ELIZABETH, 1990

PORT ELIZABETH, 1960

Bethelsdorp

INDIAN OCEAN

Motherwell

Bloemendal

Gelvandale

Malabar

INDIAN OCEAN

POPULATION

Blacks

Whites

Asians

Coloureds

City Boundary

100,000

10,000

1,000

0   5km

Feng Shui

The literal translation of *feng shui* (pronounced *fung schway*) is "wind water." Feng shui is a Chinese approach to living harmoniously with one's surrounding environment. It includes both scientific and mystical elements. Feng shui involves the correct placement of events in time (for example, picking the right day on which to get married or invest money). It also includes correctly placing objects in space. A feng shui practitioner might be asked to give advice on the location, shape, and orientation of an entire settlement or new development, an individual building, a burial site, rooms in a building, landscape elements in a garden, or furnishings in a room. The feng shui master is a type of locational consultant who uses a mix of Chinese philosophy, religion, folk cures, and science.

The roots of feng shui go back thousands of years in China. The practice has been greatly influenced by Taoism, an important Chinese philosophical system and religion. The Tao can be thought of as the path taken by natural events. Feng shui helps a person follow that path to avoid getting out of step with the world. The Tao sees the world as made up of complementary opposites called yin and yang. These opposites are constantly changing into one another in an endless cycle (think of the seasons, the water cycle, life and death, day and night). Part of the work of the feng shui master is to maintain a balance between the yin and yang of a place so that it reproduces the balance and harmony of the entire universe. Another important idea behind feng shui is that of chi, which can be thought of as spirit or energy. Chi pervades everything, and all things inhale and exhale it. To maintain harmony, beneficial chi should move smoothly and steadily through people and their surroundings.

Feng shui is a complex body of ideas and practices. There are different kinds of feng shui. What follows are just a few examples of some common feng shui rules of thumb. See how well your town, school, and classroom follow these rules.

Put mountains behind you and face south. Is your town or city placed correctly? The preferred orientation for a town and its buildings is southward-facing, with the highest mountains to the north behind it. We can understand this in light of the historical and physical geography of China. China was periodically invaded by nomadic people from the north (that is the reason the Great Wall was built). Cold winds also blew from that direction. Southward-facing mountain slopes captured the maximum amount of sunlight.

Put water in front. According to feng shui, facing a river should be beneficial. Water is associated with money and abundance in China. This belief is not surprising, since rural prosperity has come from irrigated agriculture. Rivers should be meandering and gently flowing. If they are too straight and fast, the chi associated with them will hurry by too quickly and may even turn destructive. Does your town or city face a river or a coast? If so, has this proven a good or bad location?

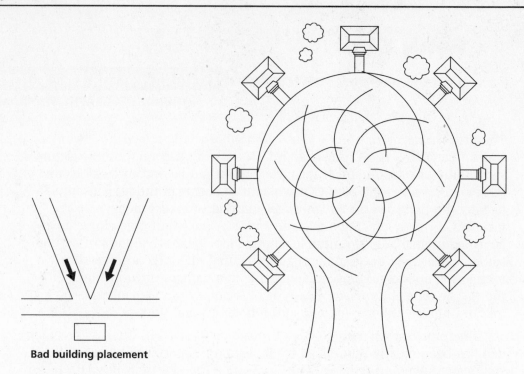

Bad building placement

Avoid a building site at the end of a street or below the confluence of two streets. For feng shui, both traffic and chi funneled at high speeds toward a building and its occupants is not healthy (see above). Likewise, a straight path leading to the front door is not as desirable as a curved one that allows the impact of any approaching harmful energy to be deflected. Therefore, sites on circular intersections or culs-de-sac are to be avoided, as chi spins around in the circular space forming a vortex. Streams of the chi can fly off and hit nearby structures, disturbing their occupants (see above). Do not position a building across from a narrow opening between two adjacent structures on the opposite side of the street. A constant stream of high-pressure energy will be aimed like a knife at such a building.

Rectangles, squares, or circles are superior to L-shaped or U-shaped spaces, where balance and the free flow of chi are harder to achieve. Round columns are preferred to square ones, as they don't have sharp, pointing angles and are less likely to interrupt the flow of chi. Long corridors, especially if both ends have doors or windows, are to be avoided, since they funnel chi too quickly through a building.

Doors facing directly across from one another along such a corridor should be kept closed to minimize potential energy conflicts. Doors are likened to parents' mouths and windows to children's. A house with too many windows or with windows that are too big or too impressive compared to the doors is bad for parental authority. By putting a bell or chime on the door, it will sound each time the door opens, thus making the window pay attention!

The most commanding spot in a room is diagonally across from the door. In a living room this spot is where the homeowner's favorite chair belongs. In an office, the most important employee should have his or her desk in this spot. It should face toward the door, allowing the occupant to see everyone who comes and goes. If there is a window in this location, then the chair or desk should be moved a bit, so the occupant's back is against a solid wall.

In cities such as Los Angeles, San Francisco, Vancouver, and New York City, feng shui consultants are well established. Check your local library for books on the subject and your local yellow pages and classified ads for feng shui practitioners. Did you find anything?

YOU ARE THE GEOGRAPHER

1. Rule 1: Put the mountains behind you and face north. How is your town oriented? Does it face in a particular direction? What advantages does it get from its orientation?

_____

_____

2. Rule 2: Avoid a building site at the end of a street or below the confluence of two streets. Does your school have any of these locational problems? What about your house? Can you name any prominent buildings in your community that do?

_____

_____

3. Rule 3: Irregular plot shapes, building shapes, and room shapes are undesirable. What is your school's shape and window configuration?

_____

_____

4. Rule 4: The most commanding spot in a room is diagonally across from the door. Where is your teacher's desk located in the room? Is it in the most commanding spot?

_____

_____

Can Korea Be United Successfully?

GEOGRAPHY FOR LIFE ACTIVITIES

NATIONAL GEOGRAPHY STANDARD 13

At the end of World War II Korea was divided into two separate countries—North Korea and South Korea. In this activity you will be given some information about North Korea and South Korea. You will then be asked to write about whether you think reunification is likely. To do that, you first need to understand what political geographers call "uniting" and "dividing" forces. Uniting forces hold a country together. Dividing forces pull a country apart. Below are some examples of these forces.

Language Language can unite or divide. The fact that practically everyone in Japan speaks Japanese ties the people and their country together. Indonesians, however, speak some 250 different languages. This is a factor that tends to divide people.

Religion Similarly, if most people in a country practice the same religion, then religion acts to unite people. If, on the other hand, the people of a country follow a variety of religious traditions, then religion may act as a dividing force. Lebanon includes several subgroups of Muslims and Christians. The country has been torn apart by religious strife for years. In nearby Saudi Arabia, however, nearly 100 percent of the people are Sunni Muslims. Here, religion is a uniting force.

Government A country is strengthened if most of its citizens have the same ideas about how their government and economy should be run. Sharply different political and economic philosophies within a country can tear it apart. For example, Germany was divided for nearly 50 years. East Germany had a communist government and a centrally controlled economy. West Germany had elected leaders. Its economy was a capitalistic, or free market, one. The reunified Germany is working hard to integrate people who were raised under these two very different systems.

Core Area Another factor is a country's core area. By core area, political geographers mean the region of a country that is heavily developed. It has many people, large cities, a dense transportation network, and lots of economic and cultural activities. Some countries have one core area. For example, nearly every French person looks to Paris as the heart of his or her country. In contrast, Spain has two core areas—Madrid and Barcelona. Spain has had to handle this situation carefully to prevent it from becoming a powerful dividing force.

History Having shared experiences binds you to your friends and family. The same is true of countries. The Swiss people are diverse when it comes to language and religious traditions, but they share a strong sense of the history of their country and its story of survival against threats from its neighbors. Belgium, however, was created by the agreement of powerful outsiders. As a result, Belgians have a weaker sense of shared history. This acts as a dividing force.

Economy Sometimes the regions of a country have very different levels of economic development. This can act as a dividing force. A rich region may feel that it can do better alone. For example, Singapore left the Malaysian

Federation (now Malaysia) to become an independent country in 1965. A poorer region may even feel it will be better able to develop on its own.

However, regions with different resources can complement each other. This can help them unite. In the case of North and South Korea, the North has most of the mineral wealth. The South, however, is better suited to agriculture.

Physical Geography A compact shape like that of Uruguay acts as a unifying force. Because various regions are close to each other and to the country's center, transportation and communications are easy. Fragmented countries like the Philippines, or long, narrow ones like Chile are harder to keep together. Mountains or seas around a country make it more easily defended. Great Britain, for example, is a country with a natural moat.

You Are the Geographer

Now look at the table on page 84. Do you think the Koreas will reunify? Write one paragraph about what you think are unifying forces. Write another paragraph about what you think are dividing forces. You may also want to look at a map of Korea and the information in your textbook.

_____

_____

_____

_____

_____

_____

_____

_____

_____

_____

_____

_____

_____

_____

_____

_____

_____

Characteristics of North and South Korea

Characteristic	North Korea	South Korea
Area	46,541 sq. mi.	38,324 sq. mi.
Population	21.9 million	47.9 million
Population density	460 per sq. mi.	1,223 per sq. mi.
Percent urban	59%	79%
Gross domestic product	$22 billion	$764.6 billion
GDP per capita	$1,000	$16,700
Value of exports	$520 million	$172.6 billion
Number of radios	4.7 million	42 million
Miles of paved highway	1,238	29,016
Life expectancy at birth	71 years	75 years
Most important ally	China	United States
Important trading partners	China, Japan, S. Korea	United States, Japan
Government type	Communist state	Republic
Economy	Centrally controlled by the government	Capitalistic (free market)
Language	Korean	Korean
Religion	Religious activities almost nonexistent under communist rule; traditionally Buddhism and Confucianism	Christianity, 49 percent Buddhism, 47 percent
Core area	P'yongyang	Seoul
	For hundreds of years before the Korean War and the partition of Korea into North and South, the Seoul region was the core area.	
Historical experiences	Koreans have been a distinct cultural group for thousands of years. The first native Korean state arose in the north in the 1st century A.D. The split between the peoples of Korea into North and South is very recent (1945).	

Name _____ Class _____ Date _____

Sacred Buildings

In many parts of the world, sacred structures are the most visible and stunning elements of the cultural landscape. In other places and traditions, houses of worship are modest. Even where there is no sacred structure, sacred places or objects are present. This activity will give you some practice identifying some famous or typical sacred structures from around the world.

Some of the structures pictured on page 87 are located in Southeast Asia. Southeast Asia has examples of Hindu, Buddhist, Islamic, and Christian sacred structures. The many kinds of religion present there are a result of the variety of cultural influences over the past 2,000 years. Today, Buddhism, Islam, and Christianity are the most widely practiced religions.

While Hinduism died out hundreds of years ago in most parts of the region, it remains alive on the Indonesian islands of Bali and Lombok. Spectacular remains of Hinduism's past importance in the form of architecture and other arts are found throughout the region.

Southern or Theravada Buddhism is the major religion of lowland Myanmar, Thailand, Laos, and Cambodia. Chinese or Mahayana Buddhism is most common in Vietnam. Indonesia is the world's most populous Muslim country. The majority of Malaysia is also Muslim. In the Philippines, most of the people are Roman Catholic.

Sacred structures from other parts of the world have been included in this activity. The ability to recognize these types of structures will help you identify images of places correctly. This ability will improve your skill of "reading" something about a culture from its landscape.

YOU ARE THE GEOGRAPHER

Now look through your textbook and consult other sources with pictures of sacred buildings from around the world as needed. Match each feature with its correct label or name and then with its location (country). Then write a brief paragraph comparing and contrasting the building styles for the different religions.

_____

_____

_____

_____

_____

Picture   Location

1. _____   _____   Ancient Egyptian Temple (Temple of Horus, Edfu)

2. _____   _____   Ancient Greek Temple (Temple of Athena Nike)

3. _____   _____   Hindu Temple (Brahmeshvara Temple, Bhuvaneshvar)

4. _____   _____   Hindu-Buddhist Temple Complex (Angkor Wat)

5. _____   _____   Buddhist Temple and Pagoda (Foguang Temple, Shanxi)

6. _____   _____   Buddhist Temple-Stupa (Ananda Temple, Pagan)

7. _____   _____   Buddhist Temple-Mountain (Borobudur)

8. _____   _____   Shinto Shrine (Izumo Shrine)

9. _____   _____   Mayan Temple (Tikal)

10. _____   _____   Synagogue (Capernaum)

11. _____   _____   Islamic Mosque (Suleymaniye Mosque)

12. _____   _____   Roman Catholic Cathedral, Gothic Style (Chartres)

13. _____   _____   Eastern Orthodox Church (St. Basil's)

14. _____   _____   Baptist Church (Delphi Falls)

a. Myanmar (Burma)   h. India
b. Cambodia   i. Indonesia
c. China   j. Japan
d. Egypt   k. Russia
e. France   l. Turkey
f. Greece   m. United States
g. Guatemala   n. Israel

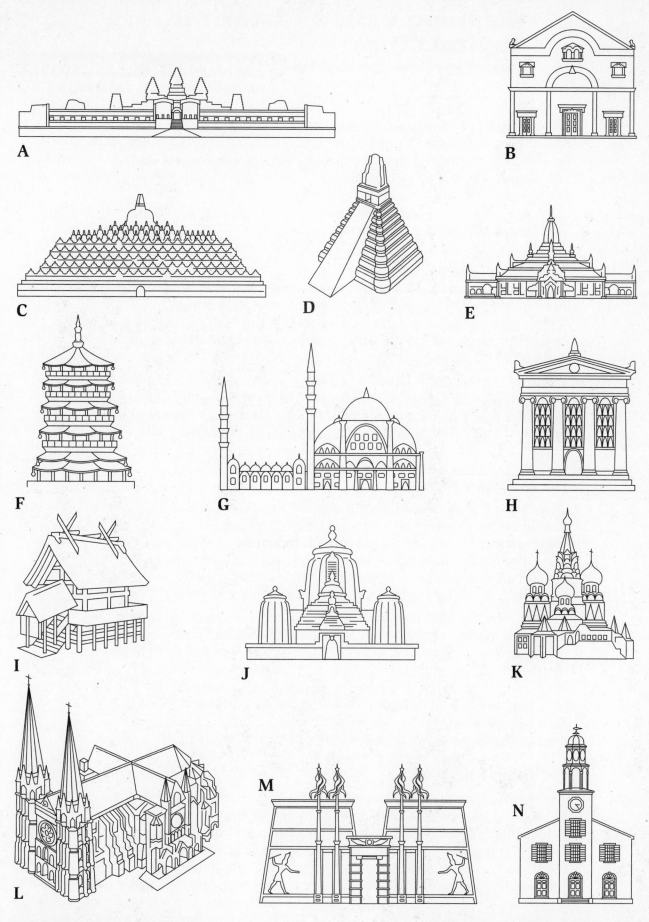

A

B

C

D

E

F

G

H

I

J

K

L

M

N

Mapping India's Historical Capital Cities

GEOGRAPHY FOR LIFE ACTIVITIES
NATIONAL GEOGRAPHY STANDARD 17

The geography of the Indian Subcontinent has changed many times. Political boundaries and capital cities have changed as various rulers and peoples have gained and then lost power. Physical features such as rivers and mountain passes as well as historical events and cultural patterns have affected the political boundaries and capital cities.

THE INDUS VALLEY CIVILIZATION

The Indus Valley civilization in modern-day Pakistan produced great cities like Mohenjo Daro and Harappa on the Indus River and its tributaries. Coming from Central Asia over the Hindu Kush, the Indo-Aryans invaded in about 1500 B.C. They spread eastward into the plain of the Ganges River and then southward into the Deccan Plateau. On the Ganges Plain, large kingdoms grew. Smaller, independent communities were more common in the southern peninsula.

By 600 B.C. northern India had at least 16 identifiable political units, each with important cities. By 400 B.C. the lands of the Ganges Plain had been formed into an empire by Chandragupta Maurya. The entire Indus Valley and lands west in modern-day Afghanistan were also in this empire. A new capital city was established at Pataliputra, now called Patna.

THE MAURYAN EMPIRE

Chandragupta's son Bindusara and grandson Asoka expanded the Mauryan Empire mainly to the south (to modern-day Bangalore). Asoka tied his empire together by building the Royal Highway (the modern-day Grand Trunk Road). This road connected his capital at Patna with many settlements along the Ganges, including the holy city of Varanasi (Benares). The road extended to the Khyber Pass, a trade and invasion route.

The centuries after Asoka's death in 232 B.C. saw many changes in the political geography of the Indian Subcontinent. Various foreign and native rulers built up and later lost kingdoms. In the A.D. 600s Sri Harsha put together a unified state in northern India that stretched from Gujarat in the west to East Bengal (modern-day Bangladesh) in the east. He was kept from the Deccan by a powerful southern dynasty. In Harsha's time, Kannauj (50 miles up the Ganges from Kanpur) displaced Pataliputra as the capital. Kannauj was sacked by the advancing Turks in 1018.

THE DELHI SULTANATE

The city of Delhi became the capital of a Turkish-Afghan dynasty in 1206. The Delhi sultan remained the most important political person in northern India until the 1500s. Delhi occupies an advantageous location where the Ganges Plain narrows between the Great Indian Desert and the foothills of the Himalaya and connects with northwest India (the Punjab). The Delhi sultanate's power reached its peak in about 1335, when most of modern-day

India's Historical Capital Cities

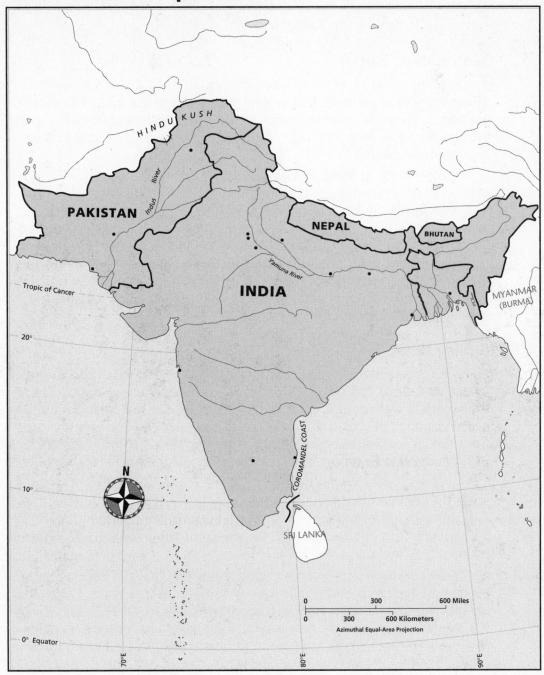

India was under its control. The sultan tried but failed to move Delhi with its entire population south to a different capital in the Deccan. Delhi was laid waste in 1398 by the Turkic invader Timur (Tamerlane).

THE MOGHUL EMPIRE

The Moghul Empire was established in India by Babur, a descendent of Timur and Genghis Khan. This empire reigned from the mid-1500s to the end of the 1600s. It was a period of great flowerings of Indian civilization. The Taj Mahal at Agra was built during this period. The Moghul emperors also beautified Delhi.

The 150 years of Moghul peace and foreign trade helped to attract foreign interest to India. Vasco da Gama was the first European to go to the area. He landed on the Malabar (southwestern) Coast of India in 1498. Soon after, the Portuguese, French, Dutch, British, and Danes established settlements. By the early 1800s the British East India Company reigned supreme. The British Crown took over the governance of India in 1858 following an Indian rebellion. Some parts of India were administered directly by Britain. Others continued as princely states governed by local rulers under treaties with the British Crown.

BRITISH INDIA

The British East India Company ran its affairs and the parts of India under its control (called "presidencies") from various headquarters cities. These headquarters were coastal cities since the purpose of the organization was overseas trade between India and Britain. Chennai (Madras) on the Coromandel (eastern) Coast, Mumbai (Bombay) on the northwestern coast, and Calcutta on the Hoogly were the seats of presidencies by the early 1700s. In 1773 Calcutta became the capital of British India.

Delhi became the capital of British India in 1912, replacing Calcutta. This move put the capital city in a more central location. This location also possessed a drier climate. In 1931 the government moved to New Delhi, a planned new town near old Delhi.

When mostly Hindu India and mostly Muslim Pakistan became independent of Britain in 1947, India kept New Delhi as its capital. Pakistan initially selected its largest city, Karachi, as its capital. In 1959 the Pakistanis built an entirely new capital, Islamabad. Its location was chosen as a means to assert Pakistan's claim to nearby Kashmir, a territory that is also claimed by India. When Bangladesh (formerly East Pakistan) separated from Pakistan in 1971, Dhaka became its national capital. It is located in the delta created by the Ganges and Brahmaputra Rivers.

YOU ARE THE GEOGRAPHER

Use maps from your textbook and an atlas to locate and label the underlined features and places on the blank outline map on page 89. After you complete your map, highlight the names of cities that have served as capitals. What can you say about the factors that influenced these choices?

Climates of the Indian Perimeter

South Asia is a land of varied climates. From a climate map, you can see that the humid tropical, tropical savanna, desert, steppe, humid subtropical, and highland climate types are all present in the region. Climatologists decide what type of climate a place has by considering its average temperatures and precipitation patterns.

YOU ARE THE GEOGRAPHER

Climagraphs are used to present detailed information about temperature and precipitation for specific locations that have weather records. The climagraphs for four cities located in the perimeter countries of the Indian Subcontinent: Karachi, Pakistan; Kathmandu, Nepal; Chittagong, Bangladesh; and Colombo, Sri Lanka are shown on page 93. Find the locations of these places on one of the maps in your textbook or in an atlas. This exercise asks you to study the information contained in the climagraphs and answer some questions that will help you understand more about South Asia and about selected climates.

1. Name the location with the highest and the one with the lowest mean (average) annual precipitation.

_____

_____

2. The distribution (rather than the amount) of precipitation throughout the year is similar for three of the stations. They have a single peak that occurs in what month? What is different about the distribution of precipitation throughout the year at Colombo?

_____

_____

3. The climates of all four stations (although much more weakly in the case of Karachi) are affected by winds that blow from the same direction for several months of the year before slackening and then reversing direction. What are these winds called?

_____

4. Look at the location of Colombo. It is on Sri Lanka, an island with water to the northeast as well as the southwest. What happens in Colombo that does not happen at the other three stations?

_____

_____

5. Which city has the least variation in maximum and minimum temperatures throughout the year? How does its latitude compare to that of the other cities? Why do places near the equator have little temperature variation throughout the year?

_____

_____

_____

6. Which station is the coolest (has the lowest mean annual maximum temperature and the lowest mean annual minimum temperature)? Explain this fact, referring to elevation, latitude, and distance from the ocean in your answer.

_____

_____

7. Rank order the four stations, beginning with the one that has the biggest range between its mean annual maximum temperature and its mean annual minimum temperature and ending with the one that has the smallest range. Can you think of reasons why temperature ranges might be greater in mountains and deserts than in humid tropical and subtropical regions?

_____

_____

_____

8. Use a climate map from your textbook or an atlas to assign each of the four cities to its proper climate type. Notice that Colombo and Kathmandu are located near boundaries between two different climates. Include both climates for these two cities.

_____

_____

_____

_____

Kathmandu, Nepal
(27° 42' N, 85° 22' E, 1336 m elevation above sea level)

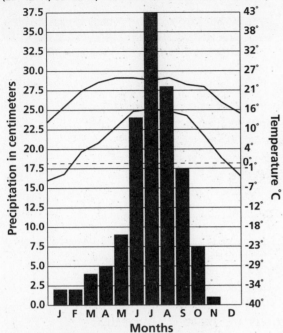

Mean annual precipitation 139.5 cm
Mean annual maximum temperature 23.8° C
Mean annual minimum temperature 11.6° C

Colombo, Sri Lanka
(8° 54' N, 79° 52' E, 7 m elevation above sea level)

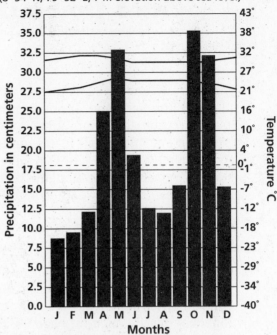

Mean annual precipitation 232 cm
Mean annual maximum temperature 30.0° C
Mean annual minimum temperature 23.9° C

Karachi, Pakistan
(24° 54' N, 67° 09' E, 22 m elevation above sea level)

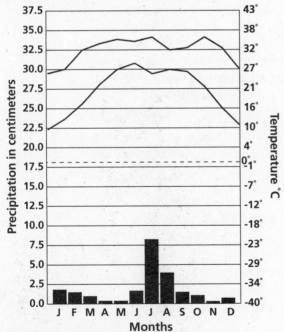

Mean annual precipitation 20.09 cm
Mean annual maximum temperature 31.6° C
Mean annual minimum temperature 20.2° C

Chittagong, Bangladesh
(22° 21' N, 91° 50' E, 27 m elevation above sea level)

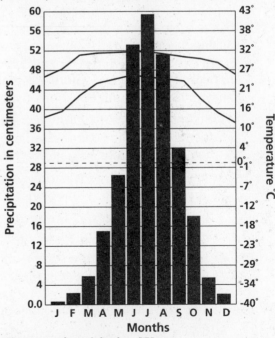

Mean annual precipitation 273 cm
Mean annual maximum temperature 29.4° C
Mean annual minimum temperature 20.6° C

The International Air Linkages of Cities Down Under

A city grows and changes mainly through its connections with other places. Cities with many connections tend to be large and prosperous. Transportation and communication linkages are the first step to economic development. Almost all international flights used to fly to and from Australia's two largest cities, Sydney and Melbourne. That pattern has started to change. The changing pattern reflects the role that each city plays within the Australian economy.

YOU ARE THE GEOGRAPHER

The origins and destinations of airlines have been grouped into seven regions: Europe, Japan and Northeast Asia, North America (including Guam), Southeast Asia, New Zealand, the South Pacific, and the Remainder (which includes South Asia, the Middle East, Africa, and South America). On the map on page 95, these regions are outlined. Trace the outlines of the regions in the following colors: Europe—green, Japan and Northeast Asia—yellow, North America—blue, Southeast Asia—orange, New Zealand—brown, South Pacific—red, the Remainder—purple.

Now look at Table 1 on page 96. Across the top are Australia's major international airports. On the map of Australia label each state and the seven cities, underlining the names of the cities that are capitals. Each number in Table 1 is a percentage. The "44" in the upper left hand corner of the table means that the Sydney airport handled 44 percent of all the airline passenger traffic between Australia and Europe in 1985. By comparing a city's figures for 1985 and 1996, you can see how its share of traffic between Australia and another region changed over that period. Now answer the following questions using Table 1 and the map.

1. In both 1985 and 1996 which city captured the largest share of international air traffic? Use the same colored pens to *draw a solid line* connecting each foreign region with this Australian city on the world map.

2. In 1996 Melbourne was the second most important Australian airport for which two regions?

3. In 1996 Brisbane was the second most important Australian airport for which three regions?

4. In 1996 Perth was the second most important Australian airport for which two regions? Using their assigned colors, *draw a dashed line* connecting each region with its second most important Australian airport—Melbourne, Brisbane, or Perth—on the world map.

5. Now use the map, Table 1, and Table 2 to make sense of the pattern of linkages. Why are certain world regions paired with certain Australian airports? People fly for three reasons: business, tourism, and social visiting. Write your thoughts down in a couple of paragraphs.

All Routes Lead to Australia

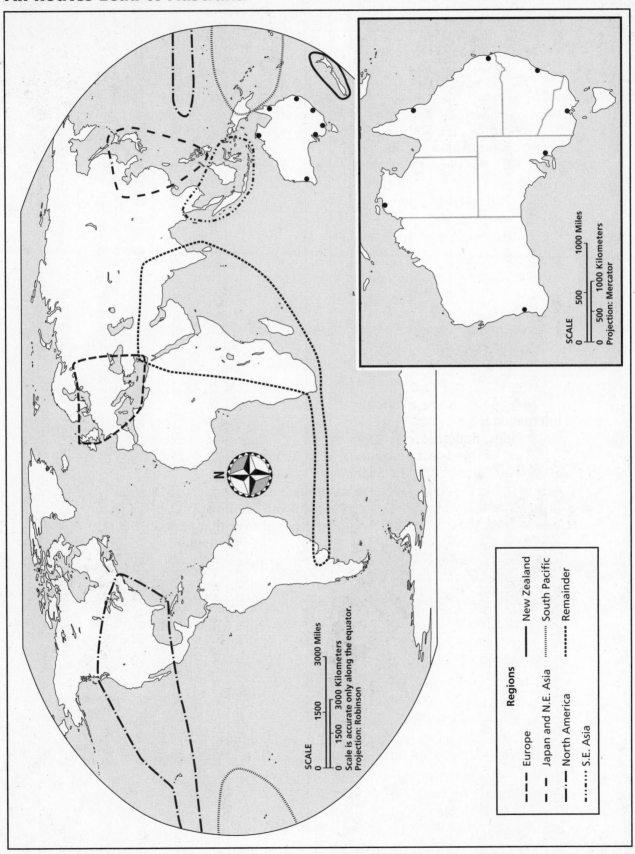

Regions

— — — Europe

— — New Zealand

— · — Japan and N.E. Asia

·········· South Pacific

—·—· North America

- - - - Remainder

·········· S.E. Asia

SCALE

0   1500   3000 Miles

0   1500   3000 Kilometers

Scale is accurate only along the equator.
Projection: Robinson

SCALE

0   500   1000 Miles

0   500   1000 Kilometers

Projection: Mercator

Table 1. Major Airports' Percentage Shares of Australian Air Traffic

Major Region	Sydney 1985	Sydney 1996	Melbourne 1985	Melbourne 1996	Brisbane 1985	Brisbane 1996	Perth 1985	Perth 1996	Adelaide 1985	Adelaide 1996	Darwin 1985	Darwin 1996	Cairns 1985	Cairns 1996
Europe	44	49	33	25	6	7	11	12	5	3	0	1	0	3
Japan & N. E. Asia	61	53	24	9	9	21	6	4	0	1	0	0	0	12
North America	69	77	22	18	4	2	0	0	0	0	0	0	5	2
Southeast Asia	34	35	27	19	4	12	23	24	4	4	3	3	0	3
New Zealand	58	50	17	20	16	26	3	2	2	0	0	0	0	1
South Pacific	62	46	14	11	19	28	0	0	0	0	0	0	4	15
Remainder	66	58	10	18	1	4	23	22	0	0	0	0	0	0
TOTAL	53	49	23	16	9	16	10	10	2	2	1	1	1	5

Percentage Share of Total Traffic

Table 2. Selected Characteristics of Australian Cities

Sydney	Population: 3.7 million, capital of New South Wales, Australia's oldest city, major commercial and industrial center, port, four universities, many historical and cultural resources, famous beaches, immigrants from all over the world
Melbourne	Population: 3.2 million, capital of Victoria, gold rush roots (1850s), commercial and industrial center, two universities, famous botanical garden
Brisbane	Population: 1.6 million, capital of Queensland, commercial and industrial center, port, three universities, subtropical climate and vegetation, anchor of Queensland's Gold Coast
Perth	Population: 1.3 million, capital of Western Australia, gold rush roots (1890s), commercial, industrial, and mining center, port, two universities
Adelaide	Population: 1.1 million, capital of South Australia, port, rail and industrial center, three universities
Darwin	Population: 78,000, capital of Northern Territory, supply and shipping point, port, city
Cairns	Population: 65,000, tourist resort, port, commercial center for the sugarcane industry, access point for the Great Barrier Reef, fishing, tropical rain forests

Diffusion Shapes the Pacific Islands

The cultures and cultural landscapes of the Pacific Islands contain elements that diffused, or spread, there from a variety of places. Humans first came to New Guinea and Australia 40,000 to 50,000 years ago from Southeast Asia, perhaps during a glacial period when the sea level was much lower than it is now. No one reached the islands further east for tens of thousands of years. It wasn't until sailing craft and navigational skills had developed that people from Southeast Asia began to arrive. By about 1500 B.C., the island groups of eastern Melanesia and western Polynesia began to be settled. Groups also moved north into Micronesia. By about A.D. 1000 the far-flung corners of the Polynesian triangle—Hawaii, Easter Island, and New Zealand, had been reached. With the exception of Papua New Guinea, the languages of Melanesia, Polynesia, and Micronesia are all members of the Austronesian (or Malayo-Polynesian) family.

Europeans did not begin arriving until the late 1700s. British Captain James Cook explored the waters around Australia and New Zealand. Europeans eventually became the majority populations of Australia and New Zealand. In the minority on most other islands of the Pacific, they became the ruling political and economic powers. The Europeans brought with them their languages, religions, political traditions, and trading economies. The colonial powers also imported workers from various parts of Asia, adding more ethnic diversity.

YOU ARE THE GEOGRAPHER

Now, study the data table provided on pages 98–99, along with a map of Oceania. Be sure to read all the notes at the bottom of the table. Then write a short essay supporting the following statement: "The Pacific Islands clearly show the effects of the diffusion of many different people and cultures into the region." This statement should be the topic sentence of your first paragraph. Evidence from the table that you can use to support your argument includes the islands' political links, ethnic groups, languages, religions, and exports.

_____

_____

_____

_____

_____

_____

_____

Selected Characteristics of Major Pacific Island Countries and Other Political Units

Political Unit	Major Ethnic Groups	Principal Languages	Principal Religions	Major Exports
MELANESIA				
Papua New Guinea	Papuan, Melan	English, 717 Papuan languages	Indig 34%, RC 22%, Lutheran 16%	gold, copper ore, oil, logs, coffee, palm oil, cocoa, lobster
Soloman Islands	Melan 93% Poly 4%	English, Melan, Poly languages	Ang 34%, RC 19%, Bap 17%, Other Chr 26%	cocoa, fish, timber, copra, palm oil
Vanuatu	Melan 94% French 4%	French, English, Bislama	Presby 37%, Ang 15%, RC 15%, Other Chr 10%, Indig 8%	copra, cocoa, coffee, frozen fish, timber, beef
New Caledonia	Melan 43% Euro 37% Poly 12% Indo 4% Viet 2%	French, Melan, Poly languages	RC 60%, Prot 30%	ferronickels, nickel ore
Fiji	Fijian (Poly) 51% Indian 44%	English, Fijian, Hindustani	Chr 52%, Hindu 38%, Muslim 8%	sugar, clothing, gold, processed fish, lumber
POLYNESIA				
Tonga	Poly	Tongan, English	Free Wesleyan 41%, RC 16%, Mormon 14%	squash, fish, vanilla, root crops, coconut oil
Samoa (W. Samoa)	Poly	Samoan, English	Chr 99.7%	copra, fish, beer, coconut oil and cream
American Samoa	Poly 93% Cauc 2% Other 6%	Samoan, English	Congr 50%, RC 20%, Prot & Other 30%	fish products
Cook Islands	Poly Cauc	English, Cook Is. Maori	Prot 70%	copra, tropical fruit, vegetables, clothing
French Polynesia (incl. Tahiti)	Poly 78% Chinese 12% French 10%	French	Prot 54%, RC 30%, Other 16%	cultured pearls, coconut products, mother-of-pearl, vanilla, shark
Wallis and Futuna	Poly	French, Wallisian	RC	copra, handicrafts
Tuvalu	Poly 96%	Tuvaluan, English	Congr 97%	copra
Hawaii	Asian & Pac Is 62%, Cauc 33%, Black 3%	English	Chr	processed sugar, canned pineapple, clothing

Table (continued)

Political Unit	Major Ethnic Groups	Principal Languages	Principal Religions	Major Exports
MICRONESIA				
Nauru	Nauruan 58%, other Pac Is 26%, Chi 8%, Euro 8%	Nauruan, English	Prot 58%, RC 24%, Confucian & Taoist 8%	phosphates
Kiribati (also partly in Polynesia)	Micro	English, I-Kiribati (Gilbertese)	RC 54%, Prot 41%	copra, seaweed, fish
Federated States of Micronesia	9 Micro & Poly groups	English, Micro & Poly languages	RC 50%, Prot 47%	fish, clothing, bananas, black pepper
Palau (Belau)	Micro	English, Palauan	RC & Prot 66%, Indig	copra, coconut oil, fish
Marshall Islands	Micro	English, Marshallese, Japanese	Chr, mostly Prot	coconut oil, fish, trochus shells
Northern Mariana Islands	Micro, Cauc, Japanese Chi, Korean	English, Chamorro, Carolinian	RC	clothing, copra, coconut oil

Abbreviations and Notes

Major Ethnic Groups: Melan = Melanesian, Micro = Micronesian, Poly = Polynesian, Euro = European, Indo = Indonesian, Viet = Vietnamese, Cauc = Caucasian, Pac Is = Pacific Islander, Chi = Chinese

Languages: Underlined languages are the official languages of independent countries. Melanesian, Micronesian, and Polynesian languages are members of the Austronesian language family, which also includes the predominant languages of Madagascar, Indonesia, and the Philippines.

Religions: Indig = Indigenous (already present when Europeans arrived), RC = Roman Catholic, Ang = Anglican, Bap = Baptist, Other Chr = other Christian, Presby = Presbyterian, Prot = Protestant, Congr = Congregational

Major Exports: Underlined exports are those introduced from outside the region to be produced locally for export. Clothing refers to European-style clothing for export. The region's largest "export" is tourism, which brings in a growing share of wealth from outside the region. Many islands in the Pacific, even those that are politically independent, still depend on foreign assistance. Tourism and foreign government subsidies as well as trade are important paths for diffusion.

Answer Key

ACTIVITY 1

1. amount of forest lost in individual South American countries between 1980 and 1990; Paraguay, Ecuador, Venezuela, Bolivia; Suriname, Guyana, Uruguay
2. Students' maps should accurately reflect the data.
3. shows countries on map, specific numbers for each country not shown; deforestation rates are now shown on a map of the countries; answers will vary
4. Answers will vary.

ACTIVITY 2

2. Graphs should accurately reflect the data.
3. June, July, August, September, October, November; September; December, January, February, March, April; June through November
4. Answers will vary; summer and fall; North America receives more solar energy at that time; it warms the oceans
5. December through March, because it should be the opposite of hurricane season in the Northern Hemisphere

ACTIVITY 3

1. The ocean currents are affected by the wind patterns.
2. 6 years; no; answers will vary
3. Answers will vary/Estimates: 1—North Pacific Ocean; 2—Sitka, Alaska; 3—British Columbia, Canada; 4—Ocean Falls, British Columbia, Canada; 5—Port Alberni, Vancouver Island, Canada; 6—Coast of Washington, U.S.; 7—Coast of Oregon, U.S.; 8—Crescent City, California, U.S.; 9—Hawaii, U.S.; 10—Philippines; 11—Coast of Oregon, U.S.
4. westerlies (and northeast trade winds); circle, oval, or ellipse; clockwise; gyre
5. Oregon; yes; 6 years; answers will vary
6. It would follow the same route as the shoes. Sea life could be killed or seriously injured. Some plants and animals might be killed by the oil spill, and this might affect others.

ACTIVITY 4

1. They all have semiarid climates.
2. San Diego, California—Urban; Hargeysa, Somalia—Grassland, grazing land; Floriano, Brazil—Grassland, grazing land; Monterrey, Mexico—Cropland, grazing land; Baghdad, Iraq—Cropland; answers will vary, but should mention that intense land use might cause loss of natural vegetation and soil erosion; answers will vary, but should mention semiarid climate, population, and land use patterns
3. Diagrams will vary.
4. Answers will vary. Answers will vary; possibly rates of erosion or plant loss, land use data, farming and grazing data; answers will vary

ACTIVITY 5

Answers will vary depending on the location of your town.

ACTIVITY 6

Students' mall maps should reflect the correct colors for the types of stores.
3. The women's apparel stores tend to be sited next to the expensive department stores.
4. electronics, sports, food, and men's apparel
5. They cluster together.
6. Answers will vary.

ACTIVITY 7

2. Students' maps should have 5 black dots, 7 brown, 8 green, 6 red, and 7 yellow. This will add up to a total of 33, not 27, because there are 6 metros that once had

NHL teams that no longer do (Quebec, Hamilton, Winnipeg, Hartford, Cleveland, and Kansas City).

3. Students maps should accurately reflect the cities of the NHL.

4. Answers will vary.

ACTIVITY 8

Essays should include a discussion of the many differences in the pioneering experience, the quality of the land, differences in the socioeconomic status of pioneers, and reasons for pioneering.

ACTIVITY 9

Students' decisions on the tourism growth in Dominica should be backed up with information or logical reasoning.

ACTIVITY 10

Students' tables should include all the relevant information included in the activity.

ACTIVITY 11

Students' maps will vary depending on how they chose to organize their data. Students may correlate a drop in GDP with poor soil areas, mountainous areas, rain forests, or due to a location with few resources.

ACTIVITY 12

1. Chile, Peru, Ecuador, Bolivia
2. Chile, Ecuador, Peru, Bolivia
3. Ecuador, 37%; Bolivia, 36%; Chile, 15%, Peru, 11%
4. Bolivia, $525; Chile, $171; Ecuador, $160; Peru, $30
5. Peru; Ecuador
6. 14%
7. bananas, cocoa beans, pineapples, rice; warm climate; it is more tropical
8. coffee, peaches, sugar; coffee; desert; with irrigation

9. cotton, soybeans; in prairie regions
10. apples, grapes, maize, milk, pears, wine; California; apples, grapes, wine

ACTIVITY 13

Students' maps should include all the underscored words in the activity.

ACTIVITY 14

Students' completed maps should accurately reflect the data given.

ACTIVITY 15

1. Denmark, Norway, Sweden
2. They went west from the Faeroe Islands.
3. They went south from Denmark and Sweden.
4. They went to eastern Europe, Russia, and Asia from Sweden.
5. The earliest date on the map is 620 in the Shetland Islands in what is now the United Kingdom.

Answers will vary for the essay but should demonstrate an understanding of the major issues involved in migration discussed in the activity.

ACTIVITY 16

4. capitals on the Danube: Sofia, Bucharest, Budapest, Belgrade, Sarajevo, Ljubljana, Zagreb, Vienna, and Bratislava
5. capitals on the Elbe: Prague, Berlin
6. capitals on the Vistula: Warsaw
7. Tirane, Skopje
8. The Elbe flows north and empties into the North Sea. Barge traffic slowed to a mere trickle after the Iron Curtain.
9. The Danube flows mainly west to east. It empties into the Black Sea. Traffic on the Rhine is greater than on the Danube because the Rhine flows through much wealthier countries in the heart of Europe.
10. The Rhine-Main-Danube Canal has increased commerce for the cities on its banks.

ACTIVITY 17

Students' maps should accurately label all the cities from the table. Answers to the essay question will vary. See map below.

ACTIVITY 18

1. cotton, vineyards, vegetables, fruit, wheat, dairying, tea, citrus fruit, tobacco, corn, silkworms; the Colchis gets more precipitation; used for irrigation **2.** livestock raising, grains, fruit, vineyards, vegetables; vines, tree crops, and animals often do well in hilly areas **3.** livestock on seasonal pastures; too cold in winter **4.** Baku, T bilisi, Gänca **5.** Answers will vary.

ACTIVITY 19

1. 1971: 96, 90, 85, 110; 1976: 90, 83, 70, 140; 1989: 73, 59, 33, 300; 1993: 69, 53, 27, 350; 1998: 66, 46, 22, 450; 2010: 59, 35, 15, 600; **2.** volume **3.** about six times saltier **4.** Graphs should reflect the information in the table.

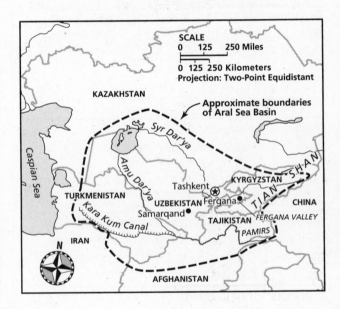

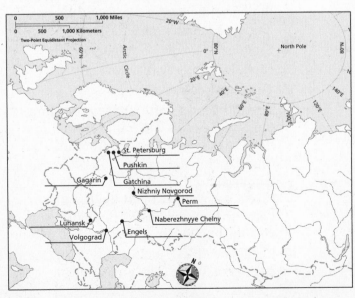

ACTIVITY 20

1. Bahrain, Kuwait, Qatar; to work or shop
2. Egypt; business, schooling
3. Sudan; jobs
4. Both countries are isolated politically.
5. Indonesia
6. to work
7. economic reasons; oil
8. Pakistan
9. Muslims who settled in South Africa during colonial times
10. Saudi statistical yearbook, world almanacs

ACTIVITY 21

Answers will vary, but students should use facts to support their choices.

ACTIVITY 22

Stories or poems will vary, but facts about the camel should be accurate.

ACTIVITY 23

Students' maps should accurately reflect the proportions of migration groups.

ACTIVITY 24

Students' drawings should show an accurate understanding of the series of events.

ACTIVITY 25

Students' essays should include pertinent details relevant to the position they have taken.

ACTIVITY 26

Students' essays should show an understanding that apartheid caused the black population to be moved into separate areas. Small black servant populations are reflected in every white area. Also, coastal areas and the city center are all white. The discussion of the comparison of American Jim Crow laws should recognize that they too represented the same basic aim. It is easier to control a population group if it is concentrated in certain manageable areas.

ACTIVITY 27

Answers will vary.

ACTIVITY 28

Students' answers may vary, but should include data from the tables to back up their answer.

ACTIVITY 29

Picture Location
1. M d
2. H f
3. J h
4. A b
5. F c
6. E a
7. C i
8. I j
9. D g
10. B n
11. G l
12. L e
13. K k
14. N m

ACTIVITY 30

Students' maps should show correct placement of labels.

ACTIVITY 31

1. highest: Chittagong; lowest: Karachi
2. July; Colombo has two peak rainy periods each year

3. monsoons
4. receives year-round precipitation
5. Colombo; much lower; receive Sun's direct rays year-round
6. Kathmandu; higher elevation, furthest north, most inland
7. Kathmandu, Karachi, Chittagong, Colombo; high elevations, distance from oceans, influence of high and low pressure systems
8. Students should correctly place climate types.

ACTIVITY 32

1. Sydney
2. Europe and North America
3. Japan and Northeast Asia, New Zealand, South Pacific
4. Southeast Asia, Remainder
5. Answers will vary.

ACTIVITY 33

Answers will vary but students' essays should demonstrate an understanding of how various cultural groups have influenced the Pacific Islands.